Secure Data Mining

Secure Data Mining

Jocelyn O. Padallan

ARCLER
PRESS

www.arclerpress.com

Secure Data Mining

Jocelyn O. Padallan

Arcler Press
224 Shoreacres Road
Burlington, ON L7L 2H2
Canada
www.arclerpress.com
Email: orders@arclereducation.com

ISBN: 978-1-77469-103-8 (Hardcover)

ABOUT THE AUTHOR

Jocelyn O. Padallan is Assistant Professor II from Laguna State Polytechnic University, Philippines and she is currently pursuing her Master of Science in Information Technology at Laguna State Polytechnic University San Pablo Campus and has Master of Arts in Education from the same University. She has passion for teaching and has been Instructor and Program Coordinator at Laguna State Polytechnic University

TABLE OF CONTENTS

LIST OF FIGURES

LIS OF TABLES

PREFACE

Every sphere of human life is burdened with a huge amount of database, and this bulky database gives rise to a need for tools powerful enough to transform this data into valuable knowledge. To meet the demands of the database, a number of ways were explored by the researchers to develop mechanisms and methods in the areas of pattern recognition, neural nets, machine learning, data visualization, statistical data analysis etc. The researchers have developed from the endeavors a new field of research, often termed as ***data mining and knowledge discovery***.

This era of information technology has a distinctive features of enormous amount of data being produced and stored by all forms human activities. Computers are used to store a huge portion of this database called computer databases, making the data accessible by the computer technology. However, enormous amount of data creates a problem of extraction of valuable knowledge from the database.

Big Data cannot be stored or handled by the conventional data storage systems and hence the analysis tools of conventional systems are not capable enough to examine big data. Additionally, storing of big data in cloud storage give rise to the challenges of the data privacy breach. There are attacks based on data mining, an unauthorized or hostile user can access to the classified data mined from the raw data through computation which generates a major threat to the data. This books gives a secure method of data mining techniques taking into account the privacy and security of the data. Even in the disturbed environment, this approach can keep up the validity and authenticity of the data to produce the data after computation.

Fundamentals and basic concepts regarding data mining are given in Chapter 1 which include data types, information gained from the data, and usefulness of the data mined. Chapter 2 provides detailed knowledge about the security of the data in the process of data mining. A number of approaches of security including classification and detection of data, clustering of data, intrusion detection systems etc. are discussed in this chapter. Classification approaches of the data are discussed in Chapter 3 of this book. Categorization of data and categorization techniques, preprocessing of data and feature selection are the presented in this chapter. Chapter 4 discusses the application of secure data mining in fraud detection. This chapter gives overview of the existing fraud detection systems and compares it with the secure system of fraud detection. The techniques used for fraud detection including Bayesian networks, Rule-based algorithms, Artificial Neural networks etc. are discussed in detail in this chapter. Application of data mining in crime detection is presented in Chapter 5 of this book. This chapter starts with the introduction of intelligent crime analysis and then gives detailed overview about the crime detection techniques used in data mining which include Self-Organizing Map Neural Network, Crime Matching etc. Chapter 6 is dedicated to the interdisciplinary nature of the data mining with telecommunication. Role of data mining in telecommunication, multidimensional association and sequential pattern analysis, use of visualization tools in telecommunication data analysis etc. are discussed in detail. Chapter 7 presents interconnection between data mining and security systems. Role of data mining in security systems and real-time data mining-based intrusion detection systems are explored in this chapter. Finally, Chapter 8 gives insight about the recent trends and future projections of data mining. A comparison of the past data mining trends with the present and future trends is given in this chapter. Interdisciplinary nature of the data mining with other fields of engineering and science, finance and retail industries is also discussed in this chapter. This book can serve as a valuable tool for the readers from diverse fields of data security along with the researchers and experts of data mining.

Author

Chapter 1

Fundamentals of Data Mining

CONTENTS

1.1. INTRODUCTION

The information age is a term used to describe our current era. We have been gathering enormous volumes of data in this information era since we think that information leads to strength and prosperity, especially due to advanced technology like satellites, computers, etc. With the introduction of computers and the ability to store large amounts of data digitally, we began gathering and storing a wide range of data, relying on the capacity of computers to sift through this jumble of data (Zhang & Wu, 2011). However, the huge amounts of data stored on different structures became unmanageable very quickly. As a result of the original turmoil, organized databases and DBMS (database management systems) have been developed. Effective DBMS have been highly essential assets for managing a big amount of information and, particularly, for efficient and effective recovery of specific data from a vast collection whenever required. The current enormous collection of all kinds of data has also been aided by the development of DBMS. From written reports, corporate transactions and scientific information to satellite images, and military intelligence, we now have considerably more data than we may manage ` is no longer sufficient for judgment. Presented with massive amounts of data, we›ve developed new requirements to assist us in making decent management decisions (Adhikari, 2012). Such requirements include automated data summarizing, retrieval of the "essence" of data stored, and pattern discovery in original information.

1.2. MOVING TOWARD THE INFORMATION ERA

The phrase "we are living in the information age" is widely used; nevertheless, we live in the age of information. Every day, petabytes or terabytes of information from society, business, medicine, research and engineering, and nearly every other facet of everyday lifestream into our computer networks, WWW (World Wide Web), and different information storing devices. The rapid development of modern data collecting and storage techniques, as well as the digitalization of our society, has resulted in an explosive rise in available data volume (Adhikari & Rao, 2008). Sales transactions, stock trading data, product specifications, sales campaigns, efficiency & profile of the company and feedback of the customer are just some of the massive data sets generated by businesses across the world. Large retailers, like Walmart, for instance, process millions of dollars of transactions every week across hundreds of locations across the globe. Remote sensing, process measurement, science studies, the efficiency of the system, engineering inspections, and environmental monitoring all produce petabytes of data regularly in engineering & science activities (Zhang et al., 2010).

Each day, worldwide core communication networks transport tens of petabytes of data. Medical records, monitoring of patients and diagnostic imaging create massive volumes of data in the health and medical industries. Hundreds of petabytes of data are processed every day as a result of billions of online queries facilitated via search engines. Social media & communities have grown in importance as the sources of data, resulting in computerized videos and images, Web communities, blogs, and many types of social networking. Several sources create massive volumes of data (Ramkumar & Srinivasan, 2008).

Our period is genuinely the information era because of the rapidly expanding, publicly available, and massive quantity of data. To automatically extract useful information from massive volumes of data and turn it into structured knowledge, efficient and adaptable technologies are desperately needed. Data mining has been born as a result of this need. The field is new, vibrant, and exciting. Data mining has made significant progress in our transition from the information era to the future data age, and it would continue to do so (Han et al., 2007).

1.3. DATA MINING

Data mining, being a genuinely multidisciplinary discipline, may be described in a variety of methods, which is unsurprising. Even the phrase "data mining" doesn't encompass all of the key elements. We use the term gold mining rather than sand or rock mining to describe the gold mining from sand or rocks. Similarly, data mining must have been renamed "knowledge mining from data," which is regrettably a bit of a mouthful (Chung & Gray, 1999; Hand, 2007).

Figure 1.1. Data mining—searching for knowledge in information.

Source: https://www.sciencedirect.com/science/article/pii/B9780123814791000010

Furthermore, data mining can't represent the focus on mining from huge volumes of data in the short term. Nonetheless, mining is a colourful phrase that describes the process of extracting a limited number of valuable nuggets from a large amount of raw material (Figure 1.1) (Romero & Ventura, 2013). As a result, a misnomer including both the words "data" and "mining" became a popular option. Several additional words, such as mining of knowledge from information, the extraction of knowledge, pattern or data analysis, data dredging, and data archaeology, have comparable meanings to the mining of data (Witten & Frank, 2002).

Several people confuse data mining with another often-used phrase, knowledge discovery from data (KDD), while others see data mining as only one stage in the process of knowledge discovery. With the massive amounts of data saved in databases, files, and other sources, it is becoming increasingly critical, if it is not required, to create sophisticated tools for data analysis and interpretation, as well as the extraction of useful information that may aid in decision-making.

The considerable excavation of implicit, unidentified, and possibly valuable information from data in databases is referred to as data mining. Although KDD and data mining are commonly used interchangeably, data mining is a component of the process of knowledge discovery. Data mining is depicted in Figure 1.2 as a stage in an adaptive process of knowledge discovery (MacQueen, 1967; Ester et al., 1996).

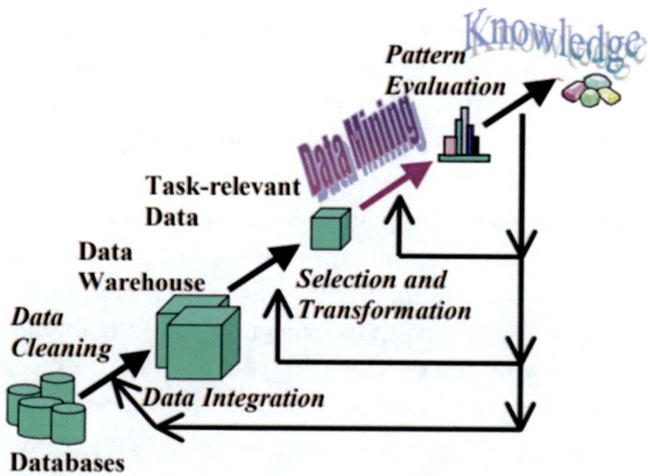

Figure 1.2. Data Mining is the core of the process of Knowledge Discovery

Source: https://webdocs.cs.ualberta.ca/~zaiane/courses/cmput690/notes/Chapter1/ch1.pdf

A few stages go from original data collection to some sort of new information in the Knowledge Discovery in the process of Databases. The phases in the repetitive approach are as follows (Freund & Schapire, 1997):

- **The cleaning of data:** often referred to as the cleansing of data, is the process of removing junk and unnecessary data from a database.
- **The integration of data:** various data sources, sometimes dissimilar, can be integrated with a single source at this step.
- **The selection of data:** this phase determines which data is important to the analysis and obtains it from the collection of data.
- **The transformation of data:** often referred to as the aggregation of data, is the process of transforming chosen data into formats that are suitable for mining.
- **The mining of data:** this is an important phase in which sophisticated approaches are used to find potentially relevant trends.

- **The estimation of pattern:** Using specified metrics, mathematically interesting patterns reflecting knowledge is determined.
- **The illustration of knowledge:** this is the last stage of the process, in which the consumer's newly obtained knowledge is graphically displayed. This crucial phase employs visualization tools to assist consumers in comprehending and interpreting the data mining findings.

A few of such stages are frequently combined. *The integration of data & the cleansing of data*, for example, may be used as part of a pre-processing step to create a database system. *The transformation of data & the selection of data* may also be coupled, with the selection resulting in data integration, or the selection is performed on changed data, as in the case of the database system (Friedman, 2001).

Iteration is a key component of the Knowledge Discovery of Data. The assessment metrics may be improved, the mining may be improved, fresh data may be picked or more processed or fresh sources of data may bc incorporated once found information has been given to the consumer, resulting in various, more relevant outcomes.

The word "data mining" comes from the resemblance between looking for useful information in a huge database and mining for lucrative metal veins in rocks. Both involve sifting through a vast quantity of data or cleverly probing the data to locate the precise location of the data (Hall et al., 2009). It is, although, a misnomer, because gold mining in rocks is typically referred to as "gold mining" rather than "rock mining," therefore data mining must have been referred to as "the mining of knowledge " alternatively. Nonetheless, data mining quickly became the recognized standard phrase, obscuring more generic words like knowledge discovery in databases, which represent a more comprehensive approach. Information extraction, data dredging, and pattern finding are other names for data mining (Breiman, 2001).

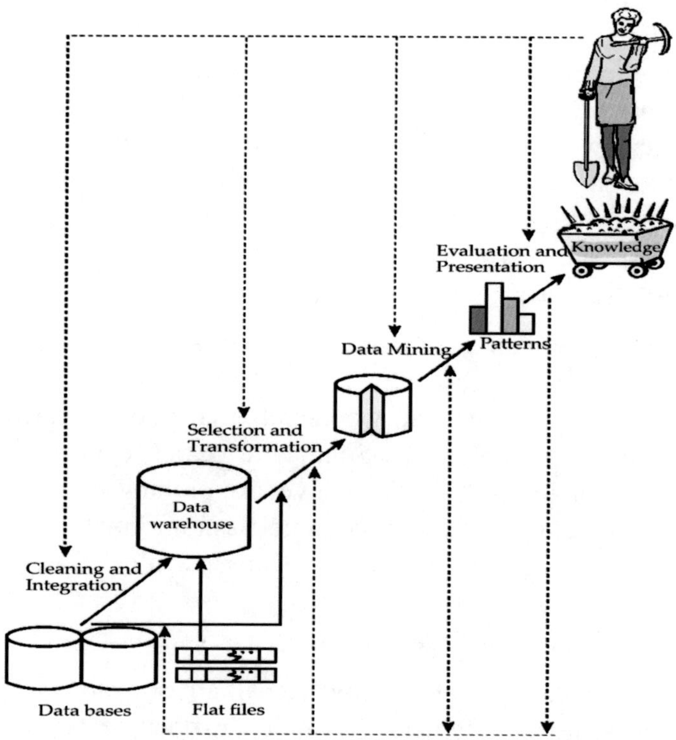

Figure 1.3. Data mining is a phase in the knowledge discovery procedure.

Source: https://www.skedsoft.com/books/data-mining-data-warehousing/introduction-to-data-mining

1.4. WHAT TYPE OF DATA ARE WE GATHERING?

We've been gathering a wide range of information, from precise mathematical measures and text files to more sophisticated data including geographical data, multimedia channels, and web texts. This is a non-exhaustive list of information stored in databases and flat files in digital format (Nazib & Moh, 2021).

1.4.1. Business Transactions

Every commercial transaction is usually "memorized" for the rest of the time. Inter-business activities, like swaps, buying, stocks, banking, etc, or intra-business activities, like administration of in-house products and resources, are examples of these transactions. Major departmental stores such as to retain millions of transactions every day, amounting to terabytes of data, owing to the extensive usage of bar codes (Novikov et al., 2019). Because the pricing of hard drives is constantly falling, storage capacity is not a big issue; however, the efficient usage of information in a realistic time frame for strategic decision making is the most critical challenge for firms struggling to survive in a fiercely competitive environment.

1.4.2. Scientific Data

Our society is accumulating colossal amounts of scientific information that require to be analyzed, either in a Switzerland nuclear accelerator laboratory counting particles, in a Canadian woodland having studied measurements from a grizzly bear radio collar, on a South Pole ice field collecting data about oceanic activity, or in an American university probing human behaviour. Regrettably, we may collect and store more fresh information quicker than we may evaluate the existing information (Jepsen et al., 2004).

1.4.3. Personal and Medical Data

Large amounts of data on groups and individuals are collected regularly, from government censuses to customers and employees documents. Governments, businesses, and institutions like hospitals are collecting vast amounts of personal information to aid in the management of human resources, market research, and customer service. Irrespective of the private problems that this sort of data raises, it is collected, utilized, and even exchanged. When combined with the other information, this information may provide insight into customer behaviour and other topics (Lacasse et al., 2012).

1.4.4. Observation of Images and Video

Video cameras are more common because the price of video cameras continues to drop. Observation camera's videos and photos are frequently reprocessed, and the material is damaged as a result. Nowadays, although, there is a trend to retain and even digitize the recordings for future usage and study (Thurston, 2004).

1.4.5. Satellite Sensing

Many satellites are circling the Earth, some of which are geostationary over an area and others are circling the Earth, however, all of which are constantly providing data to the surface. National Aeronautics and Space

Administration (NASA), which is in charge of a huge number of satellites, receives more data per second than all of the National Aeronautics and Space Administration engineers & researchers may handle. Several satellite images and information are available quickly as feasible with the aim of being analyzed by other scientists (Takaishi et al., 2014).

1.4.6. Games

Our civilization accumulates a vast quantity of information and analytics about sports, athletes and players. All data is saved, including basketball passes, hockey scores, and car-racing gaps, as well as swimming timings, chess positions, and boxer's pushes. Such data is being used by journalists and commentators for coverage, although athletes and trainers will like to use it to enhance efficiency and gain a better understanding of their rivals (Mayer et al., 2014).

1.4.7. Digital Media

One of the reasons for the growth of digital technology storage is the development of low-cost scanners, digital cameras, and portable cameras. Several radio stations, TV channels, and movie studios are also digitizing their sound and visual archives to effectively manage their multimedia resources. Organizations like the NBA & the NHL already have begun transferring their massive game records to digital formats (Webster & Ksiazek, 2012).

1.4.8. Software Engineering and Computer Assisted Design Data

CAD (Computer Assisted Design) systems are used by designers to design buildings and engineers to design system circuits & components. Such systems generate a massive amount of information. Furthermore, software engineering generates a lot of comparable data, such as function libraries, objects, code, etc that requires the use of maintenance and strong administration tools (Lethbridge et al., 2003).

1.4.9. Virtual Worlds

The 3-D virtual locations are utilized for various uses. Specific languages, like VRML, are used to explain such areas and the things they include. Such virtual locations should ideally be specified in a way that allows them to exchange locations & objects. There are a lot of virtual reality items and location resources to choose from. Although the quantity of the archives continues to expand, administration of such sources, and also material search and retrieval from such collections, are ongoing research challenges (Dede, 1992).

1.4.10. Memoranda and Text Reports

The majority of interactions inside and between corporations, research groups, and even private individuals are dependent on text documents and memoranda, which are frequently sent via electronic mail. Such communications are saved in digital format regularly for future usage and reference development in strong digital libraries (Alexander et al., 1996).

1.4.11. The World Wide Web Repositories

The files of various forms, descriptions, and content have been gathered and interlinked via hyperlinks as the World Wide Web's debut in 1993, making it the greatest storehouse of data ever constructed. Due to the wide range of topics encompassed and the limitless accomplishments of publishers and assets, the Www is a very essential collection of data routinely utilized for reference, notwithstanding its vibrant and unorganized nature, non-homogenous characteristics, and frequent duplication and lack of consistency. Several people believe that the World Wide Web would eventually become a storehouse of human thought (Cooley et al., 1996).

1.5. WHAT KIND OF DATA CAN BE MINED?

Theoretically, data mining is not limited to a single form of data or media. Every type of data storage must be considered for data mining. When applied to various kinds of information, however, methods and methodologies can vary. Furthermore, the problems raised through various forms of data are vastly different. Databases, like object-oriented databases, object-relational databases, and relational databases, transnational databases, data warehouses, semi-organized and unorganized resources like the WWW, sophisticated databases like multimedia databases, spatial databases, textual databases, and time-series databases, and even flat files are all being used and researched with data mining. Here are several specific instances (Macskassy, 2007):

1.5.1. Flat files

Flat files, particularly at the laboratory scale, are the most frequent source of data for data mining techniques. Flat files are basic textual or binary data files having a design that the data mining system can recognize. Transactions, time-series data, scientific assessments, and other types of data may be found in such files (Sharma & Mehta, 2012).

1.5.2. Relational Databases

In a nutshell, a relational database is a collection of tables that hold the values of object characteristics or the values of characteristics from object relationships. Tables include columns & rows, with rows representing tuples and columns representing characteristics. In a relational table, a tuple represents an item or a connection among entities and is identifiable by a collection of characteristic values that show a primary key. We show some Items, Consumer, and Borrow connections in Figure 1.4, which illustrate business activities in a hypothetical video store called OurVideoStore (Han et al., 1996). Such relationships are only a subset of what may be a database for a video shop, and they are provided like an illustration.

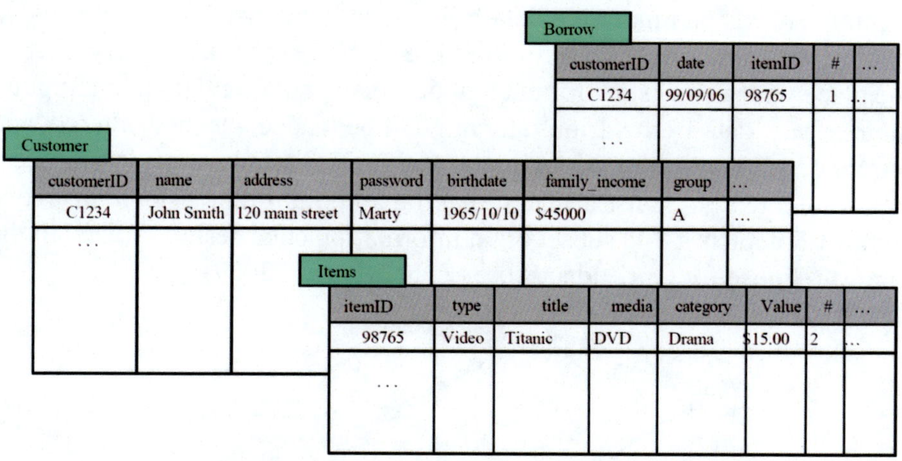

Figure 1.4. A relational database's fragments of specific relations.

Source: https://www.researchgate.net/figure/Fragments-of-some-relations-from-a-relational-database-for-OurVideoStore_fig2_242778793

SQL is the most common query language for relational databases, and it permits for the access and corruption of information contained in tables, and the computation of aggregation functions like mean, summation, minimum, maximum, and counting. An SQL query to select videos sorted by genre, for example, might be (Houtsma & Swami, 1995):

SELECT count(*) FROM Items WHERE type=video GROUP BY category.

Because they may avail the benefit of the design inherent in relational databases, data mining techniques based on relational databases may be more adaptable than data mining techniques based on flat files. Although SQL may help with the selection of data, conversion, and aggregation, data mining extends beyond what SQL might give, like forecasting, comparing, and identifying discrepancies (Mehenni & Moussaoui, 2012).

1.5.3. Data Warehouses

A data warehouse, sometimes known as a storehouse, is a collection of data from many resources that are meant to be utilized overall under the same central schema. A data warehouse allows you to evaluate data from many resources all under one roof. Let's say OurVideoStore expands to North America as a franchise. Many of the video stores owned by OurVidcoStore can have distinct databases and architectures (Lyman et al., 2008). If a business executive needs permission to data from across all outlets for good decision making, strategic direction, promotion, etc, it's best to have everything in one place with a consistent format that permits dynamic evaluation. To put it another way, data from various storage will be loaded, cleaned, altered, and combined. A multi-dimensional data design is typically used to represent data warehouses to assist decision-making and multi-dimensional displays. Figure 1.5 depicts a 3-D subset of an information cube design utilized in the data warehouse for Our Video Store (Zubcoff et al., 2009).

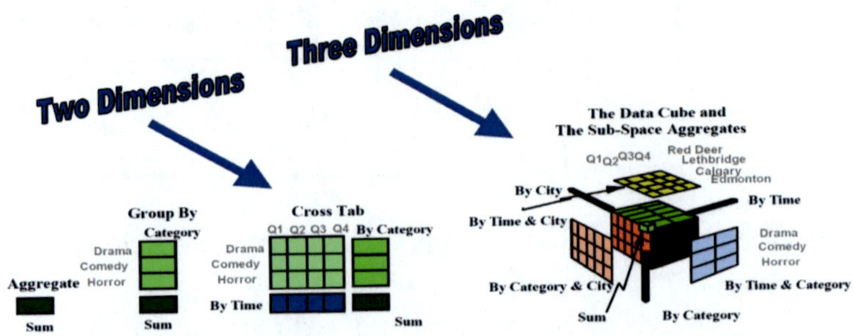

Figure 1.5. A multi-dimensional data cube form that is frequently utilised in data for data warehousing

Source: http://webdocs.cs.ualberta.ca/~zaiane/courses/cmput690/notes/Chapter1/

The figure presents a cross table of summarized rentals by film time & categories, followed by a summary of rents by film time & categories (in quarters). The data cube summarizes the rents in 3 components: time, category, and location. A cube is made up of cells that hold the values of certain standard measures (such as rental counts) as well as unique cells that hold summaries across dimensions (Inmon, 1996). A pyramid of values for one characteristic exists in every dimension of the data cube.

Data cubes are highly adapted for quick dynamic searching and assessment of information at multiple conceptual levels, termed as OLAP (On-Line Analytical Processing), due to their design, the pre-computed summary data they include, and the hierarchical attribute values of their dimensions. Roll-up, drill-down, dice, slice, and other On-Line Analytical Processing procedures enable users to navigate data at multiple levels of analysis (Palpanas, 2000). The roll-up (on the location aspect) and the drill-down (on the time horizon) actions are depicted in Figure 1.6.

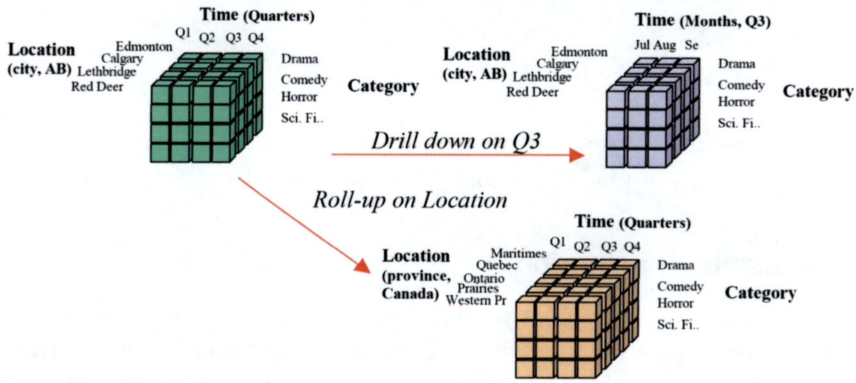

Figure 1.6. Recapitulated data after and before roll-up and drill-down procedures

Source: https://www.researchgate.net/figure/Summarized-data-from-OurVideoStore-before-and-after-drill-down-and-roll-up-operations_fig1_262367677

1.5.4. Transaction Databases

A transaction database is a collection of documents that each has an identifier, a timestamp, and a list of objects. There may additionally be descriptive information about the objects linked with the transaction documents. In the context of a video shop, the transaction database is represented by the rentals table, as illustrated in Figure 1.5. Every record is a rental agreement that includes customer identification, a date, and a catalogue of rented goods (such as VCRs, videotapes, games, etc.). Transactions are generally kept in flat files or 2 normalised transaction tables, First for the transactions and 2nd for the transaction items, because relational databases don't support stacked tables (e.g. a collection as attribute value) (Chen et al., 1996). The so-called market basket analysis or rules of linkage, in which relationships amongst items that appear together or in succession are investigated, is a common data mining research on these datasets.

Rentals

transactionID	date	time	customerID	itemList
T12345	99/09/06	19:38	C1234	{I2, I6, I10, I45 …}
...				

Figure 1.7. A portion of a rental transaction database.

Source: https://www.researchgate.net/figure/Fragment-of-a-transaction-data-base-for-the-rentals-at-OurVideoStore_fig3_242778793

1.5.5. Multimedia Databases

Audio, video, pictures, and text material are all included in multimedia databases. They may be kept on file systems, enhanced object-oriented databases. Multimedia has a higher degree of complexity that makes data mining much more difficult. Computer graphics, computer vision, picture analysis, and natural language processing methods can be required for data mining using multimedia sources (Zaiane et al., 1998).

1.5.6. Spatial Databases

Spatial databases are archives that contain geographic data such as maps and national or international locations in addition to regular data. Data mining techniques face additional problems when dealing with geographical datasets (Han et al., 1997).

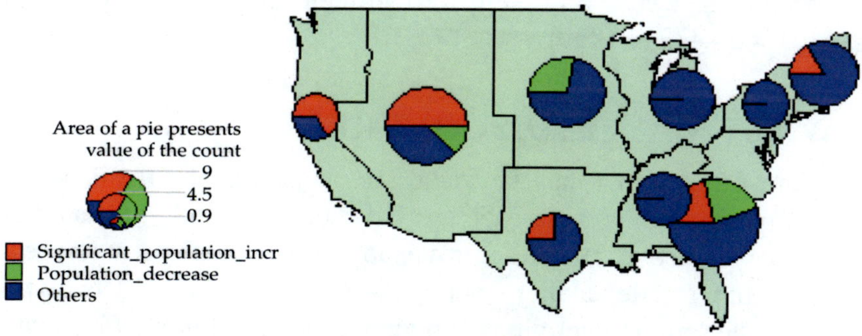

Figure 1.8. The idea of spatial On-Line Analytical Processing

Source: https://www.slideserve.com/denim/spatial-data-mining-and-spatial-data-warehousing-special-topics-in-database

1.5.7. Time-Series Databases

Time-series databases store data that changes over time, like a stock market record or recorded actions. Such databases often receive a steady stream of fresh data, necessitating the need for time-consuming real-time research. The analysis of patterns and connections between the evolutions of various factors, and the forecast of patterns and movements of the factors through time, are frequent applications of data mining in these kinds of databases (Last et al., 2001).

1.5.8. World Wide Web

The World Wide Web is the most vibrant and diverse archive accessible. A vast number of writers and publication companies are constantly participating in its development and transformation, and a huge number of consumers utilize its services daily. The content on the World Wide Web is structured into papers that are linked together. Video, text, raw data, music, and even programmers may be used to create such files (Mughal, 2018). The World Wide Web is made up of 3 key components: The web's data which includes the records accessible the web's design, which includes hyperlinks and content linkages; and the web's use, which describes when and how the sources are utilized. A 4th dimension, referring to the papers' dynamic nature, might be introduced. Data mining or web mining, on the World Wide Web, attempts to solve all of such difficulties and is typically separated into three categories: web use mining, web structure mining, and web content mining (Cooley et al., 1999).

1.6. WHAT MAY BE DISCOVERED?

The types of trends that may be found are determined by the data mining activities performed. Generally, there have been 2 kinds of data mining techniques: predictive and descriptive. Predictive data mining jobs seek to make estimates depending upon judgment on existing data, whereas descriptive data mining activities explain the basic features of current data (Fayyad et al., 1996).

The following is a quick overview of data mining capabilities and the types of information they investigate:

1.6.1. Categorization

Data categorization is the summary of fundamental characteristics of objects in a target class that results in feature rules. A database enquiry is used to obtain data related to a consumer-specified class, which is then processed via a summary module to retrieve the core of the data at multiple levels of analysis. Consumers that borrow more than thirty movies per year, for instance, can be described as OurVideoStore clients. The *attribute oriented induction* technique may be utilized to do the summary of data using idea hierarchy on the attributes explaining the target class (Lehmann et al., 2005). Basic On-Line Analytical Processing procedures meet the aim of data categorization with a data cube including the summary of data.

1.6.2. Discrimination

Data discrimination is the association of generic characteristics of items among 2 classes known as the target class and the opposing class, resulting in discriminating criteria. For instance, one could wish to compare the basic features of consumers who leased over thirty movies in the previous year to those who leased below five. Data discrimination methods are essentially the same as data categorization methods, with the distinction that data discrimination findings contain comparable measurements (Ruggieri et al., 2010).

1.6.3. Association analysis

The identification of what are known as the rules of association is the goal of the analysis of association. It evaluates the frequency of items appearing together in transactional databases and finds large items using a resource dependency. Another criterion utilized to localize association rules is a certainty, which is the conditional likelihood that an object will occur in a transaction when another object does. Market basket analysis frequently employs association analysis. For instance, knowing which films are frequently leased together or whether there is a correlation between renting a specific sort of film and purchasing snacks might be helpful information for the OurVideoStore management. P→Q [s,c], where Q and P are attributed value-pair conjunctions, s (used for support) is the probability that Q and P exist together in a transaction, and c (used for confidence) is the conditional probability that Q exists in a transaction if P is available. The hypothetic association rule, for instance (Tan et al., 2004):

Rent Type(X, "game") ∧ Age(X, "13-19") → Buys(X, "pop") [s=2%,c=55%] this means that two per cent of the transactions examined are from consumers aged thirteen to nineteen, who are leasing a game and purchasing snacks, and that certainty of fifty-five per cent exists that young consumers who rent a game also purchase snacks.

1.6.4. Categorization

The grouping of data into classes is referred to as categorization analysis. The categorization, also known as supervised categorization, employs supplied class labels to arrange the items in the collection of data. In most categorization techniques, all items are previously linked with predefined class labels in a training dataset. The categorization method develops a model by learning from the training dataset (Lehmann et al., 2005). Fresh items are classified using the model. For instance, after implementing payment terms, OurVideoStore management may examine client's responses concerning payment and classify clients who got rewards with one of 3 labels: "extremely risky", "dangerous," and "safe". The categorization study will result in a model that may be utilized in the future to approve or disapprove the requests for money.

1.6.5. Prediction

Due to the possible consequences of effective forecasting in a commercial environment, the forecast has gotten a lot of attention. Forecasts may be of 2 kinds: first, may attempt to anticipate inaccessible data values or upcoming trends, or one may attempt to anticipate a class label for the certain dataset. The second is connected to categorization. The class label of an item may be predicted depending upon the attribute values of the item and the attribute values of the classes after a categorization model is created depending upon a training dataset. Forecasting is more commonly known as the forecasting of deficient numeric value or the decreased/increased patterns in a time-related dataset (Soni et al.., 2011). The main concept is to evaluate a vast number of historical values to predict likely future values.

1.6.6. Clustering

Clustering is the grouping of data into classes, comparable to categorization. In contrast to categorization, class labels are undefined in clustering, and it is up to the clustering method to find appropriate classes. Since this categorization is not limited by supplied class labels, clustering is also known as *unsupervised categorization* (Ng & Han, 2002). There are a variety of clustering techniques, all of which are dependent upon the idea of increasing intra-class resemblance and reducing resemblance among items of various classes (*inter-class resemblance*) (Bharara et al., 2018).

1.6.7. Outlier analysis

Data items that may not be classified into a single class or cluster are known as outliers. They're also referred to as surprises or outliers, and they're crucial to spot. Although outliers may be labelled noise in certain uses and disregarded, they may disclose crucial information in other areas, making them extremely relevant and its analysis is beneficial (Ilango et al., 2012).

1.6.8. Evolution and deviation analysis

The examination of time-related information that varies through time is called evolution and deviation analysis. Evolution analysis is a type of data analysis that allows you to characterize, compare, categorize, or group time-related data. Deviation analysis looks at the discrepancies among predicted and measured values and tries to figure out what's causing the variances from the expected values.

Customers regularly lack a clear understanding of the types of trends they may or must find from the data available. As a result, having a dynamic and comprehensive data mining system that permits the finding of various types of knowledge at various levels of analysis is critical. Interactivity is thus a key feature of a data mining system (Chamatkar & Butey, 2014).

1.7. IS EVERYTHING YOU DISCOVER FASCINATING AND BENEFICIAL?

Data mining allows for the finding of previously unrecognized and possibly beneficial information. It is highly subjective to determine if the knowledge obtained is fresh, helpful, or entertaining, and it is dependent on the utilization and the client. Data mining has the capability of generating or discovering a huge amount of trends or rules. The number of regulations may be in the millions in certain situations. A meta-mining step may also be considered to analyze the larger data mining findings. To decrease the number of trends or rules identified with a high likelihood of being uninteresting, the patterns must be measured (Marwick, 2014). This presents the issue of fullness. The client will like to learn all of the principles or trends, but only those that are important to them. The evaluation of how fascinating a finding is, known as *interactiveness*, may be dependent on measurable objective aspects like the patterns' validity when evaluated on fresh data with a high degree of certainty, or subjective descriptions like the patterns' *novelty*, *utility*, or *understandability*.

Patterns discovered may also be fascinating if they support or verify a theory that needs to be validated, or whether they suddenly challenge a well-held view. This raises the challenge of expressing what is interesting to find, like meta-rule guided discovery, which defines types of rules before the discovery phase, and interestingness modification languages, which interactively interrogate the results after the discovery stage for interesting trends. In most cases, user-defined criteria are used to determine how fascinating something is. The completeness of the patterns identified is defined by such criteria (Feyel, 2002).

For the assessment of mined information and the knowledge discovery from data procedure as a whole, assessing and identifying the interestingness of rules and patterns identified, or still to be found, is critical. However, certain tangible metrics exist, determining the value of newly found information remains a major research challenge.

1.8. DIFFERENT TYPES OF DATA MINING SETUPS

There are a plethora of data mining solutions on the market or in development. Certain systems are confined to a specific data resource or have restricted data mining capabilities, while others are more adaptable and thorough. Data mining systems may be classified using several different metrics, including the following (Rajkumar & Reena, 2010; Darabad et al., 2015):

- **Classification based on the kind of data resource extracted:** such categorization divides data mining systems into categories based on the types of data they manage, like geographic data, multimedia data, time-series data, text data, and the World Wide Web.
- **Data model categorization:** such categorization divides data mining systems into categories dependent upon the data model used, like relational databases, object-oriented databases, data warehouses, transactional databases, etc.
- **Categorization based on the type of knowledge discovered:** Such categorization divides data mining systems into categories depending on the type of knowledge obtained or data mining capabilities like discrimination, characterization, linkage, clustering, categorization, and so on. Certain systems are more extensive than others, combining various data mining functions.
- **Categorization based on the mining methods utilized:** Various approaches are used and provided by data mining systems. Such categorization divides data mining systems into categories based on the data analysis method employed, like neural networks, learning of machine, visualization, genetic process, statistics, database, and so on. Interactive exploratory setups, autonomous setups, and Query-driven systems may all be classified according to the level of client engagement engaged in the data mining procedure. A sophisticated setup will provide a wide range of data mining methods to meet various scenarios and alternatives, as well as varying levels of user engagement.

1.9. PROBLEMS IN DATA MINING

Data mining techniques are collection of approaches that have been around for a long time but have just recently been deployed as dependable and adaptable tools that consistently beat earlier statistical techniques. Although data mining is in its adolescence, it is quickly becoming a popular practice. Several unresolved difficulties must be resolved before data mining becomes a well-established, mature, and trustworthy profession. A few of these concerns are discussed further below. It's worth noting that such problems aren't exclusive and aren't arranged in any particular sequence (Yang & Wu, 2006).

1.9.1. Social and Security Challenges

Every collection of data that is exchanged and designed to bc utilized for sound decision making must consider security. Furthermore, vast volumes of sensitive and confidential data regarding organizations or persons are acquired and kept when information is collected for consumer insights, client behaviour analysis, connecting personal information with other data, and so on. Due to the sensitive nature of some of this information and the possibility of unauthorized access, it becomes contentious. Furthermore, data mining might reveal new knowledge acquisition about people or organizations, which might violate privacy laws, particularly if the data is potentially shared (Seifert, 2004). A further problem that comes as a result of such concerns is the proper application of data mining. Due to the extreme value of data, datasets of all kinds of topics are frequently sold, and because of the competitive edge that may be gained from knowledge acquisition found, a few vital data may be excluded, whereas other data may be widely disseminated and utilized without restriction.

1.9.2. Problems with the User Interface

Data mining techniques can be beneficial if the knowledge they find is engaging and, most all, comprehensible to the client. Effective visualization tool makes it easier to analyze data mining findings and helps clients' better grasp their requirements. The capability to observe data in a proper visual style helps with a lot of data exploration analysis activities. For efficient data graphical display, there are several visual concepts and recommendations (Alcalá-Fdez et al., 2009). Furthermore, considerable work remains to be done to develop appropriate visualization tools for the huge database that may be utilized to show and modify mined data. "Screen real-estate," "data presentation," and " data engagement" are the three most important challenges in visualization and user interfaces. Inter-connectivity with information and data mining outcomes is essential because it allows the client to concentrate and improve mining activities and also see the found knowledge from various perspectives and conceptual degrees.

1.9.3. Problems during Mining Methodology

Such problems are related to the data mining techniques used and their limits. The adaptability of mining algorithms, the variety of data accessible, the dimensional domain, the wide analysis requires (if defined), the evaluation of the knowledge found, the manipulation of prior knowledge and metadata, the influence and managing of distortion in the information, and other factors may all influence mining technique selections (Anand et al., 1998). Keeping a range of data mining techniques accessible, for instance, is usually advantageous because various methods can perform differently based on the data. Moreover, various approaches can fit and satisfy the needs of the client in various methods.

The majority of techniques presume that the data is noise-free. O obviously, this is a bold assumption. Exceptions, incorrect or missing data, and other factors in most databases can complicated, if not obfuscate, the analysis procedure and, in several circumstances, jeopardize the quality of the results. As a result, data preparation (cleaning and modification) becomes increasingly important. Data cleansing is frequently viewed as a waste of time, yet it is one of the most essential steps in the knowledge discovery procedure, despite how time-consuming and unpleasant it can be (Sinha, 2013). Data mining strategies must be able to deal with noise in the data as well as incomplete information.

For data mining methods, the size of the search area is even more important than that of the size of the information. The number of parameters in the domain space frequently determines the size of the search area. When the number of parameters is increased, the search area generally expands rapidly. This is referred to as the dimensional problem. This "problem" has such a negative impact on the efficiency of various data mining methods that it has become one of the most pressing challenges to address (Shahbaz et al., 2010).

1.9.4. Efficiency Problems

For information processing and explanation, a variety of artificial statistical and intelligence approaches are available. Such approaches, on the other hand, were frequently not built for the massive data sets that data mining now deals with. The size of a terabyte is rather frequent. This expresses concern about the sustainability and effectiveness of data mining approaches when dealing with enormous amounts of data. Data mining techniques with exponentially or even medium-order polynomial complexity are not viable. The most common type of technique is a linear one. Similarly, rather than mining the entire dataset, sampling may be utilized. Furthermore, issues like specimen selection and completion can occur. Parallel programming and progressive upgrading are two more performance-related concepts (Cen et al., 2007). If the database may be segmented and the findings may be combined afterwards, concurrency may undoubtedly assist with the size issue. Whenever fresh data becomes accessible, progressive upgrading is useful for combining the findings from concurrent mining or upgrading data mining findings without having to re-analyze the whole database.

1.9.5. Information Source Problems

There are several difficulties relating to information resources, certain of which are realistic, like data type variety, and others which are philosophic, as the data glut issue. Researchers have an abundance of information because we already have more than we may manage and continue to gather data at an even quicker speed. If the widespread use of dataset management setups has aided in the collection of information, the emergence of data mining is undoubtedly promoting the even greater collection of data (Chen & Liu, 2004). The present approach is to gather as much information as feasible right now and analyze it (or attempt to understand it) afterwards. The question is that we'll be gathering the correct information in the appropriate quantity, whether we identify what we're doing with it, and if we may tell the difference between crucial and inconsequential information. There's the topic of heterogeneous datasets and the focus on different complicated information types when it comes to sustainable data resource concerns. Distinct kinds of information are stored in various repositories. It is impossible to anticipate a data mining setup to obtain better mining findings upon all types of information and resources effectively and efficiently. Various methods and approaches can be needed for various types of information and resources (Shin et al., 2000). Nowadays, the concentration is on relational datasets and database systems, however new techniques for other kinds of complicated information must be developed. The universal data mining software for all types of information may not be feasible. Furthermore, the growth of diverse data resources, both structurally and semantically, offers significant problems to both the datasets and data mining communities (Melli et al., 2006).

REFERENCES

1. Adhikari, A. (2012). Synthesizing global exceptional patterns in different data sources. *Journal of Intelligent Systems*, *21*(3), 293-323.
2. Adhikari, A., & Rao, P. R. (2008). Synthesizing heavy association rules from different real data sources. *Pattern Recognition Letters*, *29*(1), 59-71.
3. Ahmad, P., Qamar, S., & Rizvi, S. Q. A. (2015). Techniques of data mining in healthcare: a review. *International Journal of Computer Applications*, *120*(15) 1-17.
4. Alcalá-Fdez, J., Sanchez, L., Garcia, S., del Jesus, M. J., Ventura, S., Garrell, J. M., ... & Herrera, F. (2009). KEEL: a software tool to assess evolutionary algorithms for data mining problems. *Soft Computing*, *13*(3), 307-318.
5. Alexander, R. B., Ludtke, A. S., Fitzgerald, K. K., & Schertz, T. L. (1996). Data from selected US Geological Survey national stream water-quality monitoring networks (WQN) on CD-ROM. *US Geological Survey Open-File Report*, *96*, 337.
6. Anand, S. S., Patrick, A. R., Hughes, J. G., & Bell, D. A. (1998). A data mining methodology for cross-sales. *Knowledge-based systems*, *10*(7), 449-461.
7. Anuradha, A., & Varma, G. S. (2016). PBI2D-Priority Based Intelligent Imbalanced Data Classification of Health Care data with Missing Values. *i-Manager's Journal on Computer Science*, *4*(1), 34.
8. Bal, M., Bal, Y., & Demirhan, A. (2011). Creating competitive advantage by using data mining technique as an innovative method for decision making process in business. *International Journal of Online Marketing (IJOM)*, *1*(3), 38-45.
9. Beel, D., & Wallace, C. (2020). Gathering together: social capital, cultural capital and the value of cultural heritage in a digital age. *Social & Cultural Geography*, *21*(5), 697-717.
10. Bharara, S., Sabitha, S., & Bansal, A. (2018). Application of learning analytics using clustering data Mining for Students' disposition analysis. *Education and Information Technologies*, *23*(2), 957-984.
11. Breiman, L. (2001). Random forests. *Machine learning*, *45*(1), 5-32.
12. Cen, H., Koedinger, K. R., & Junker, B. (2007). Is Over Practice Necessary?-Improving Learning Efficiency with the Cognitive Tutor

through Educational Data Mining. *Frontiers in artificial intelligence and applications*, *158*, 511.

13. Chamatkar, A. J., & Butey, P. K. (2014). Importance of data mining with different types of data applications and challenging areas. *Journal of Engineering Research and Applications*, *4*(5), 38-41.
14. Chen, M. S., Han, J., & Yu, P. S. (1996). Data mining: an overview from a database perspective. *IEEE Transactions on Knowledge and data Engineering*, *8*(6), 866-883.
15. Chen, S. Y., & Liu, X. (2004). The contribution of data mining to information science. *Journal of Information Science*, *30*(6), 550-558.
16. Chung, H. M., & Gray, P. (1999). Data mining. *Journal of management information systems*, *16*(1), 11-16.
17. Cooley, R., Mobasher, B., & Srivastava, J. (1999). Data preparation for mining world wide web browsing patterns. *Knowledge and information systems*, *1*(1), 5-32.
18. Danubianu, M. I. R. E. L. A., & Hapenciuc, V. C. (2008). Improving customer relationship management in hotel industry by data mining techniques. *Annals of the University of Craiova, Economic Sciences Series*, *7*(36), 3261-3268.
19. Daoxuan, D., Xiangdong, W., Jihuang, H., & Yuqing, G. (1993). Total-current-spectroscopy study of the electron states of clean and hydrogen-chemisorbed GaP (1.1.1.) 1× 1 surfaces. *Surface science*
20. Darabad, V. P., Vakilian, M., Blackburn, T. R., & Phung, B. T. (2015). An efficient PD data mining method for power transformer defect models using SOM technique. *International Journal of Electrical Power & Energy Systems*, *71*, 373-382.
21. Dede, C. J. (1992). The future of multimedia: Bridging to virtual worlds. *Educational Technology*, *32*(5), 54-60.
22. Ester, M., Kriegel, H. P., Sander, J., & Xu, X. (1996). A density-based algorithm for discovering clusters in large spatial databases with noise, 96(34), 226-231).
23. Fayyad, U. M., Piatetsky-Shapiro, G., & Smyth, P. (1996). Knowledge Discovery and Data Mining: Towards a Unifying Framework, 96, 82-88.
24. Feyel, F. (2002). Some new technics regarding the parallelisation of zébulon, an object oriented finite element code for structural

mechanics. *ESAIM: Mathematical Modelling and Numerical Analysis*, *36*(5), 923-935.

25. Freund, Y., & Schapire, R. E. (1997). A decision-theoretic generalization of on-line learning and an application to boosting. *Journal of computer and system sciences*, *55*(1), 119-139.
26. Friedman, J. H. (2001). Greedy function approximation: a gradient boosting machine. *Annals of statistics*, 1189-1232.
27. Hall, M., Frank, E., Holmes, G., Pfahringer, B., Reutemann, P., & Witten, I. H. (2009). The WEKA data mining software: an update. *ACM SIGKDD explorations newsletter*, *11*(1), 10-18.
28. Han, J., Cheng, H., Xin, D., & Yan, X. (2007). Frequent pattern mining: current status and future directions. *Data mining and knowledge discovery*, *15*(1), 55-86.
29. Han, J., Fu, Y., Wang, W., Koperski, K., & Zaiane, O. (1996). DMQL: A data mining query language for relational databases, 96, 27-34.
30. Han, J., Koperski, K., & Stefanovic, N. (1997). GeoMiner: a system prototype for spatial data mining. *AcM sIGMoD Record*, *26*(2), 553-556.
31. Hand, D. J. (2007). Principles of data mining. *Drug safety*, *30*(7), 621-622.
32. Hormozi, A. M., & Giles, S. (2004). Data mining: A competitive weapon for banking and retail industries. *Information systems management*, *21*(2), 62-71.
33. Houtsma, M., & Swami, A. (1995). Set-oriented data mining in relational databases. *Data & Knowledge Engineering*, *17*(3), 245-262.
34. Ilango, V., Subramanian, R., & Vasudevan, V. (2012). A five step procedure for outlier analysis in data mining. *European Journal of Scientific Research*, *75*(3), 327-339.
35. Inmon, W. H. (1996). The data warehouse and data mining. *Communications of the ACM*, *39*(11), 49-51.
36. Jepsen, E. T., Seiden, P., Ingwersen, P., Björneborn, L., & Borlund, P. (2004). Characteristics of scientific Web publications: Preliminary data gathering and analysis. *Journal of the American Society for Information Science and Technology*, *55*(14), 1239-1249.
37. Jothi, N., & Husain, W. (2015). Data mining in healthcare–a review. *Procedia computer science*, *72*, 306-313.

38. Koh, H. C., & Tan, G. (2011). Data mining applications in healthcare. *Journal of healthcare information management*, *19*(2), 65.
39. Lacasse, M., Théorêt, J., Skalenda, P., & Lee, S. (2012). Challenging learning situations in medical education: Innovative and structured tools for assessment, educational diagnosis, and intervention. Part 1: history or data gathering. *Canadian Family Physician*, *58*(4), 481-484.
40. Last, M., Klein, Y., & Kandel, A. (2001). Knowledge discovery in time series databases. *IEEE Transactions on Systems, Man, and Cybernetics, Part B (Cybernetics)*, *31*(1), 160-169.
41. Lehmann, T. M., Güld, M. O., Deselaers, T., Keysers, D., Schubert, H., Spitzer, K., ... & Wein, B. B. (2005). Automatic categorization of medical images for content-based retrieval and data mining. *Computerized Medical Imaging and Graphics*, *29*(2-3), 143-155.
42. Lethbridge, T. C., Singer, J., & Forward, A. (2003). How software engineers use documentation: The state of the practice. *IEEE software*, *20*(6), 35-39.
43. Lyman, J. A., Scully, K., & Harrison Jr, J. H. (2008). The development of health care data warehouses to support data mining. *Clinics in laboratory medicine*, *28*(1), 55-71.
44. MacQueen, J. (1967). Some methods for classification and analysis of multivariate observations. In *Proceedings of the fifth Berkeley symposium on mathematical statistics and probability*, 1(14), 281-297.
45. Macskassy, S. A. (2007). Improving learning in networked data by combining explicit and mined links, 22, 590-595.
46. Marwick, A. (2014). How your data are being deeply mined. *The New York Review of Books*, *9, 1-9*.
47. Mayer, I., Bekebrede, G., Harteveld, C., Warmelink, H., Zhou, Q., Van Ruijven, T., ... & Wenzler, I. (2014). The research and evaluation of serious games: Toward a comprehensive methodology. *British journal of educational technology*, *45*(3), 502-527.
48. Mehenni, T., & Moussaoui, A. (2012). Data mining from multiple heterogeneous relational databases using decision tree classification. *Pattern Recognition Letters*, *33*(13), 1768-1775.
49. Melli, G., Zaïane, O. R., & Kitts, B. (2006). Introduction to the special issue on successful real-world data mining applications. *ACM SIGKDD Explorations Newsletter*, *8*(1), 1-2.

50. Mughal, M. J. H. (2018). Data mining: Web data mining techniques, tools and algorithms: An overview. *Information Retrieval*, *9*(6).
51. Nandagopal, M. R., & Gopalakrishna, H. V. (1971). Impulse-Voltage Breakdown Characteristics of Large Gaps at Low Pressures. *Journal of Applied Physics*, *42*(13), 5874-5876.
52. Nazib, R. A., & Moh, S. (2021). Sink-Type-Dependent Data-Gathering Frameworks in Wireless Sensor Networks: A Comparative Study. *Sensors*, *21*(8), 2829.
53. Nemati, H. R., & Barko, C. D. (2001). Issues in organizational data mining: a survey of current practices. *Journal of data warehousing*, *6*(1), 25-36.
54. Ng, R. T., & Han, J. (2002). CLARANS: A method for clustering objects for spatial data mining. *IEEE transactions on knowledge and data engineering*, *14*(5), 1003-1016.
55. Novikov, S., Kazakov, O., Kulagina, N., & Ivanov, M. (2019). Organization of data gathering and preparing on the basis of blockchain for the supporting system of making decisions in the sphere of developing human capital of region. In *IOP Conference Series: Materials Science and Engineering*, 497(1), 012046.
56. Palpanas, T. (2000). Knowledge discovery in data warehouses. *ACM Sigmod Record*, *29*(3), 88-100.
57. Rajkumar, A., & Reena, G. S. (2010). Diagnosis of heart disease using datamining algorithm. *Global journal of computer science and technology*, *10*(10), 38-43.
58. Ramkumar, T., & Srinivasan, R. (2008). Modified algorithms for synthesizing high-frequency rules from different data sources. *Knowledge and information systems*, *17*(3), 313-334.
59. Romero, C., & Ventura, S. (2013). Data mining in education. *Wiley Interdisciplinary Reviews: Data Mining and Knowledge Discovery*, *3*(1), 12-27.
60. Ruggieri, S., Pedreschi, D., & Turini, F. (2010). Data mining for discrimination discovery. *ACM Transactions on Knowledge Discovery from Data (TKDD)*, *4*(2), 1-40.
61. Saive, R., Emmer, H., Chen, C. T., Zhang, C., Honsberg, C., & Atwater, H. (2018). Study of the interface in a GaP/Si heterojunction solar cell. *IEEE Journal of Photovoltaics*, *8*(6), 1568-1576.

62. Seifert, J. W. (2004). Data mining and the search for security: Challenges for connecting the dots and databases. *Government Information Quarterly*, *21*(4), 461-480.
63. Shahbaz, M., Masood, S. A., Shaheen, M., & Khan, A. (2010). Data mining methodology in perspective of manufacturing databases. *Journal of American Science*, *6*(11), 999-1012.
64. Sharma, L., & Mehta, N. (2012). Data mining techniques: A tool for knowledge management system in agriculture. *International Journal of scientific and technology research*, *1*(5), 67-73.
65. Sharma, N., & Dubey, S. K. (2012). A hand to hand taxonomical survey on web mining. *International Journal of Computer Applications*, *60*(3), 4-10.
66. Shin, C. K., Yun, U. T., Kim, H. K., & Park, S. C. (2000). A hybrid approach of neural network and memory-based learning to data mining. *IEEE Transactions on Neural Networks*, *11*(3), 637-646.
67. Siddiqui, T., & Ahmad, A. (2018). Data mining tools and techniques for mining software repositories: A systematic review. *Big Data Analytics*, 717-726.
68. Singh, M., Sharma, M., & Tobschall, H. J. (2005). Weathering of the Ganga alluvial plain, northern India: implications from fluvial geochemistry of the Gomati River. *Applied Geochemistry*, *20*(1), 1-21.
69. Sinha, P. (2013). Multivariate polynomial regression in data mining: methodology, problems and solutions. *International Journal of Scientific and Engineering Research*, *4*(12), 962-965.
70. Soni, J., Ansari, U., Sharma, D., & Soni, S. (2011). Predictive data mining for medical diagnosis: An overview of heart disease prediction. *International Journal of Computer Applications*, *17*(8), 43-48.
71. Takaishi, D., Nishiyama, H., Kato, N., & Miura, R. (2014). Toward energy efficient big data gathering in densely distributed sensor networks. *IEEE transactions on emerging topics in computing*, *2*(3), 388-397.
72. Tan, P. N., Kumar, V., & Srivastava, J. (2004). Selecting the right objective measure for association analysis. *Information Systems*, *29*(4), 293-313.

73. Thurston, A. (2004). Promoting multicultural education in the primary classroom: broadband videoconferencing facilities and digital video. *Computers & Education*, *43*(1-2), 165-177.

74. Webster, J. G., & Ksiazek, T. B. (2012). The dynamics of audience fragmentation: Public attention in an age of digital media. *Journal of communication*, *62*(1), 39-56.

75. Witten, I. H., & Frank, E. (2002). Data mining: practical machine learning tools and techniques with Java implementations. *Acm Sigmod Record*, *31*(1), 76-77.

76. Yang, Q., & Wu, X. (2006). 10 challenging problems in data mining research. *International Journal of Information Technology & Decision Making*, *5*(04), 597-604.

77. Yoo, I., Alafaireet, P., Marinov, M., Pena-Hernandez, K., Gopidi, R., Chang, J. F., & Hua, L. (2012). Data mining in healthcare and biomedicine: a survey of the literature. *Journal of medical systems*, *36*(4), 2431-2448.

78. Zaiane, O. R., Han, J., Li, Z. N., Chee, S. H., & Chiang, J. Y. (1998). Multimediaminer: a system prototype for multimedia data mining. *ACM SIGMOD Record*, *27*(2), 581-583.

79. Zhang, S., & Wu, X. (2011). Fundamentals of association rules in data mining and knowledge discovery. *Wiley Interdisciplinary Reviews: Data Mining and Knowledge Discovery*, *1*(2), 97-116.

80. Zhang, S., Jin, Z., & Lu, J. (2010). Summary queries for frequent itemsets mining. *Journal of Systems and Software*, *83*(3), 405-411.

81. Zubcoff, J., Pardillo, J., & Trujillo, J. (2009). A UML profile for the conceptual modelling of data-mining with time-series in data warehouses. *Information and Software Technology*, *51*(6), 977-992

Chapter 2

Security in Data Mining

CONTENTS

2.1. INTRODUCTION

In the perspective of computers, security refers to a system's capacity to safeguard information and data, as well as its sources, in such a way that a third party cannot compromise authenticity, confidentiality, or integrity (Sunar et al., 2006). Confidentiality guarantees that even a third party cannot read or comprehend the material, although Integrity prevents a third party from changing or modifying the content in its entirety or in portions. If a person is determined to be unauthorized, the authenticity function will prevent him from using, viewing, or modifying the material or resource (Yeh et al., 2015).

Intrusion refers to activities that jeopardize the confidentiality, accessibility, or integrity of one or even more computer sources. The use of filtering router settings and firewalls to combat intrusions fails to halt these assaults. Despite all efforts to create secured systems, still intrusions occur, and they must be identified as soon as they occur. Utilizing data mining methods, an intrusion detection system (IDS) can uncover persistent system pattern attributes that can be used to detect abnormalities and recognized intrusions to use an appropriate collection of classifiers (Ali et al., 2012).

Intrusion may be easily identified utilizing certain simple data mining methods such as Clustering and Classification. Classification methods are utilized to label and analyze test data into recognized forms of classes, whereas Clustering methods are utilized to collect objects into several groups so that all equivalent objects are becoming members of the identical cluster and other objects get to be participants of other clusters. Although data mining allows for the retrieval of underlying information or hidden patterns from massive amounts of data, it can also offer security risks (Vasanthakumar et al., 2015).

The goal of Privacy-Preserving Data Mining (PPDM) is to protect sensitive content from unintentional or unauthorized disclosure (Cai et al., 2011). So far, several PPDM techniques have been suggested. Depending on their enforcement of the privacy rule, a few of these are shown in Fig. 2.1.

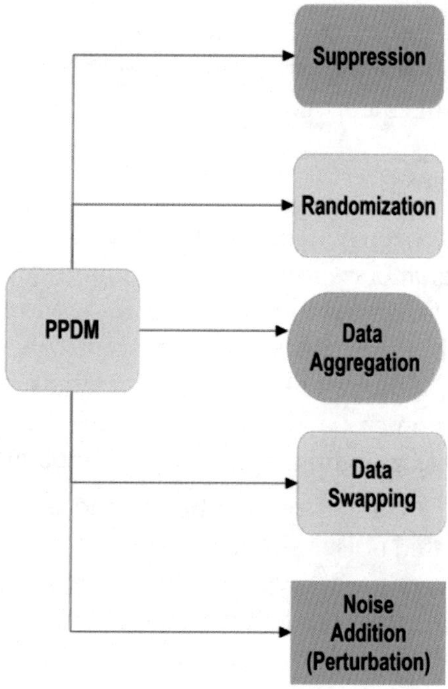

Figure 2.1. Privacy-Preserving Data Mining Techniques:

Source: https://globaljournals.org/GJCST_Volume16/6-Security-in-Data-Mining.pdf

2.1.1. Suppression

Before each computation, all sensitive or private information around an individual, like address, name, salary, age, and other data, is obscured. Rounding (Rs/- 35462.33 may well be adjusted to 35,000), Generalization (Name Louis Philip may well be substituted with the letters LP, and Location Hamburg may well be substituted with HMG, and so on) are among the tactics used to obscure this information. Suppression, on the other hand, cannot be employed if data mining demands full accessibility to sensitive information. Instead of hiding the sensitive data included inside a record, another method of suppression is to limit the identification connection of a record. DeIdentification is the name given to this method. However, one de-identification approach is k-Anonymity. It guarantees that the information published is protected from re-identification of the people to whom the data pertains (Aggarwal, 2005). It's impossible to enforce k-anonymity until all information is gathered in one trusted location. Alternatively, a cryptographic method relying on Shamir's Secret Sharing mechanism is used; unfortunately, this incurs computing overhead.

2.1.2. Randomization

Considering the existence of a corporation's central server, which collects data from many consumers and uses data mining methods to develop an Aggregate Model, the randomization enables consumers to contribute regulated noise or arbitrarily disrupt the data, removing actual information. The multiplication or addition of the numbers created arbitrarily can be used to introduce noise in numerous ways (Cai et al., 2011). Perturbation aids the Randomization process in maintaining essential anonymity.

Single records are created by combining the actual information with such randomized created noise. Single records will not be able to recoup the noise that has been introduced, giving in the required privacy. The following steps are usually included in randomization methods:

- The Data Suppliers only offer their information to the Information Receiver after randomizing it.
- A Distribution Reconstruction Algorithm is used by the information receiver to calculate the distribution.

2.1.3. Data Aggregation

In an attempt to make data analysis easier, data aggregation algorithms collect data from numerous sources. An attacker could use this to identify individual-level and private data and detect the party. Whenever the data analyst can recognize specific individuals using the mined data, his privacy is regarded to be in jeopardy. Data can be anonymized soon after aggregation to protect it from being identifiable. But, anonymized sets of data may still contain sufficient information to allow for individual identification (Lu et al., 2012).

2.1.4. Data Swapping

Just for privacy, the information swapping method involves exchanging values between distinct records. Data privacy can be retained without affecting the totals of the lower order data, enabling aggregate computations to be conducted as previously. Because this approach does not require randomization, so it can be combined with various frameworks like k-anonymity without breaking the model's privacy definitions.

2.1.5. Noise Addition/Perturbation

Differential privacy, which incorporates controlled noise, is a strategy that improves query accuracy while reducing the likelihood of data being identified (Dwork et al., 2016). The following are some of the strategies that have been employed in this regard:

- Composition in Parallel
- The Mechanism of Laplace
- Composition in Sequence

2.2. CLASSIFICATION AND DETECTION USING DATA MINING TECHNIQUES

Worms are malicious computer programs that reproduce by themselves to propagate from one machine to another. computer viruses, worms, keyloggers, spyware, Trojan horses, and adware are examples of malware. Other harmful programs include Remote to Local Worms, UDP worms, User to Root Worms, http worms, and port scan worms . Attackers create these programs for a variety of objectives, including disrupting computer processes, obtaining sensitive information, and obtaining access to private systems (Friedman et al., 2008). Catching a worm on the internet is critical because it generates weak places in the system and affects its effectiveness. As a result, it's critical to spot the worm early on and categorize it by utilizing data mining classification techniques until it produces any harm. Bayesian,

Decision Tree, Random Forest, and others are some of the categorization methods which can be used (Wu et al., 2008). The core idea of most worm detection systems is the Intusion Detection System (IDS). Because it's difficult to foresee what shape an upcoming worm will acquire, therefore automatic detection is difficult. Host-based IDS and Network-based IDS are the two forms of IDS available. The Intrusion Detection System based on the network represents network packets before approaching the end-host, whereas the Intusion Detection System based on host represents packets that have previously reached the end-host. Furthermore, detection investigations based on host encode network packets so the internet worm's stroke can be struck. If we concentrate on network packets in the absence of encoding, then we have to evaluate the network's traffic performance. In the realm of worm and intrusion detection systems, many machine learning algorithms have been employed. As a result, Data Mining and especially Machine Learning methods play a critical role in worm identification systems. Several novel strategies for building Intrusion Detection models have indeed been developed utilizing different Data Mining techniques. Machine Learning algorithms such as Genetic and Decision Trees Algorithms could be utilized to discover normal and abnormal patterns from the classifiers and a training set can then be produced relying on the test data to identify them as Abnormal or Normal classes. The existence of an intrusion can be indicated by the data marked as abnormal (Siddiqui et al., 2009).

2.2.1. Decision Trees

Among the most prevalent machine learning approaches is Quinlan's decision tree approach. The divide-and-conquer approach is used to build the tree, which consists of several leaves and decision nodes. Every decision node tests a criterion among the input data's properties and can have a variety of branches, each handling a different test outcome. A leaf node can be used to reflect the outcome of a decision. T is a collection of n-classes C1, C2,..., Cn in a training set. When T contains cases from a single class, it is referred to as a leaf. When T is blank and has no cases, it would still be considered a leaf, as well as the parent node's main class, which is assigned to the associated class (Dwork et al., 2016). If T has several classes, T is divided into k subsets (T1, T2,..., Tk), with k indicating the number of test results. The method is repeated for each Tj, whereby 1 <= j<= n since every subset is assigned to a particular class. When building the decision tree, selecting the correct characteristic for every decision node is critical. For this, the C4.5DT uses the Criterion of Gain Ratio. This criterion is used to select a property that gives the most information gain while also reducing test bias. The tree that has been created can be used to categorize test data that has similar characteristics as the training data. Starting with the root node, every testing is conducted. Each of the branches getting to a child is pursued depend on the results. The process is continued indefinitely so long as the offspring would not be a leaf. The test scenario being analyzed is provided the class as well as its appropriate leaf node (Siddiqui et al., 2009).

2.2.2. Genetic Algorithms (GA)

Genetic Algorithms (GA) are indeed a type of machine learning technique to problem-solving that uses biological evolution processes. It can be utilized to efficiently optimize a group of potential solutions. GA makes use of chromosome-like data architectures that are susceptible to evolution employing genetic operators such as mutation, selection, and crossover (Abadeh et al., 2007). Throughout the start, a chromosomes population is created at random. All alternative solutions to a problem are included in the population, which are referred to as candidate solutions. Genes are encoded as letters, numbers orbits that represent various places on a chromosome. Depends on the intended solution, a function named Fitness Function analyzes the usefulness of every chromosome. The Crossover operator represents biological reproduction, whereas the Mutation operator represents species mutation. The operator selects the most compatible chromosomes (Desale et al., 2015). Figure 2 2. illustrates how Genetic Algorithms work. The following three aspects must be addressed prior to utilizing GA to solve different issues:

- The role of fitness
- Representation of individuals
- GA Parameters

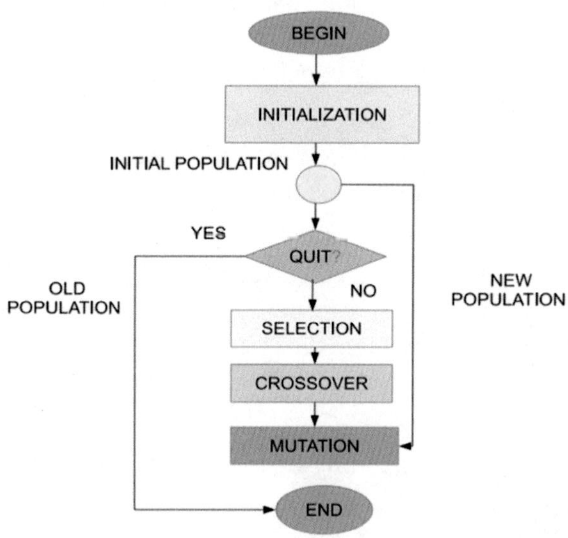

Figure 2.2. Flowchart for a Genetic Algorithm (GA)

Source: https://www.mdpi.com/2073-8994/12/11/1758/htm

Artificial Immune Systems can be designed using a GA-based technique. Wu et al. (2014) suggested a technique for smartphone malware identification based on this methodology, in which dynamic characteristics of malware are retrieved and hazardous scores of analyzed specimens are calculated.

2.2.3. Random Forest

The Method of Random Forest is a classification method formed up of a group of tree-structured classifiers which determines the winner class dependent on the votes given by the forest individual trees. Every tree is built by randomly selecting data from a training dataset. Test and Training sets can be created from the given dataset. The training set is made up of the majority of the data, whereas the test set is made up of the remainder (Abadeh et al., 2007). The steps for building a tree are as follows:

- If indeed the training dataset contains N cases, a choice of N cases from the data set is randomly chosen. The sample correlates to the training set for the tree's growth.
- A random selection of m variables from the M input variables is made, and the node is divided depending on the optimal split upon that m value. The consistent m is maintained throughout the forest's growth.
- Every tree throughout the forest is developed to its full potential. Individual trees do not receive any Pruning or trimming.
- The random forest is created by combining all of the classification trees that have been created thus far. It is frequently referred to be an operational data mining method because it can solve the issue of overfitting huge datasets and train/test complex data sets quickly.

Every classification tree is unique, and each class is voted on. Ultimately, a solution class is created depending on the total number of votes.

2.2.4. Association Rule Mining (ARM)

In datasets, association-rule mining uncovers intriguing relationships among a collection of attributes (Vittapu et al., 2015). Associative rules can be used to represent the datasets and their relationships. This data can be utilized to make decisions about a variety of activities, like shelf management and promotional pricing. Rule mining of Traditional association entails giving data analyst datasets from many firms to uncover association or patterns rules among them. Since we can perform complex analysis on these enormously big datasets in an effective cost manner, so the data owner's sensitive information can be discovered by the dataminer, which creates a privacy risk. Association rule mining is still one of the most popular pattern discovery approaches in KDD today.

To solve an ARM issue, you must traverse the objects in a database, which could be performed utilizing a variety of algorithms based on the need (Cheng et al., 2020). Depending on the strategy employed to traverse the search area, ARM techniques are typically classified as DFS (Depth First Search) or BFS (Breadth-First Search) methods. Depending on how well the support numbers for the itemsets are derived, the DFS and BFS approaches are furthermore divided into Intersecting and Counting. The Apriori-DIC, Apriori, and Apriori-TID algorithms all use BFS with Counting approaches, whereas the Partition method utilizes BFS with Intersecting techniques. On the other side, ECLAT is built on DFS with Intersecting, whereas the FP-Growth method depends on DFS with Counting techniques Such methods can be tweaked to increase the speed (Giannotti et al., 2012).

2.2.4.1. BFS with Counting Occurences

The Apriori method is the most prevalent in this category. It takes advantage of an item set's downward closure feature by trimming contenders with rare subsets before counting their followers. Confidence and Support are the two measures to examine while analyzing association rules. By understanding the support numbers of the candidates all subsets before, BFS provides the needed optimization. The enhanced computational cost of rule extraction from such a massive database is a drawback of this method. The Fast Distributed Mining(FDM) method is an insecure, modified, and distributed version of the Apriori method. Organizations may now use data extremely efficiently due to advances in data mining methods.

A single scanning of the complete database counts the contenders of a cardinality k in Apriori. The most important aspect of the Apriori Method is glancing at candidates for every transaction. A structure of hashtree is utilized for this purpose (Kumar et al., 2013). In contrast to regular Apriori, which depends on a raw database, Apriori-TID depicts every transaction depending on the present candidates it includes. The advantages of both Apriori-TID and Apriori are combined in Apriori-Hybrid. Apriori-DIC, a variant of Apriori, attempts to blur the distinction among counting, processes, and candidate creation. A prefix tree is used to do this.

2.2.4.2. BFS with Intersections

A Partition Method is analogous to the Apriori method in that it determines support values based on intersections instead of counting events occurring. The division of itemsets may lead to an exponential rise of intermediate outcomes that exceed physical memory limits. This difficulty can be solved by dividing the database into tiny portions and treating each one separately. A chunk's size is chosen such that all the lists of intermediate can be stored in memory. Optionally, a further scan can be run to confirm that now the itemsets are not just regionally but also worldwide prevalent (Giannotti et al., 2012).

2.2.4.3. DFS with Counting Occurences

A database scan is performed for each acceptable-sized contender set in Counting. The basic combination of Counting Occurences and DFS is virtually meaningless due to the computational complexity involved in scanning databases. FP-Growth, on either hand, employs the FP-Tree, a greatly condensed version of transaction data. DFS and Counting occurrences are used to create an FP-Tree.

2.2.4.4. DFS with Intersections

To pick agreeable numbers, the method ECLAT couples DFS with list intersections. It employs a technique known as Fast Intersections for optimization. It will not necessitate the breaking up of the database because the entire classes path starting from the root will be kept in memory. The procedure of mining association rules grows faster because this method removes almost all of the computational complexity (Kumar et al., 2013).

2.3. CLUSTERING

Clustering is among the most extensively utilized data mining finding approaches. It enables you to cluster a data set so that the intra-cluster resemblance is maximized whereas the inter-cluster resemblance is minimized (Herman et al., 2012). Clustering entails the uncontrolled learning of a large number of unknown classes. The clustering methods can be divided into various categories, as shown in Fig. 2.3.

- Based on the centroid
- Clustering based on connections or hierarchical clustering

- Based on a distribution
- Modern Clustering Techniques and others.
- Based on Density

2.3.1. Connection Based Clustering

Clustering depending on connections (hierarchical) is built on the assumption that objects are much more connected to each other than to distant things. The distance in between components is taken into account by the Connection Based Clustering methods while connecting them to create clusters. Rather than a single dataset segmentation, these techniques give an extended structure of combining clusters at specific distances. Clusters are represented using a Dendrogram. Its y-axis depicts the clusters' merging distance (Shapiro, 2012).

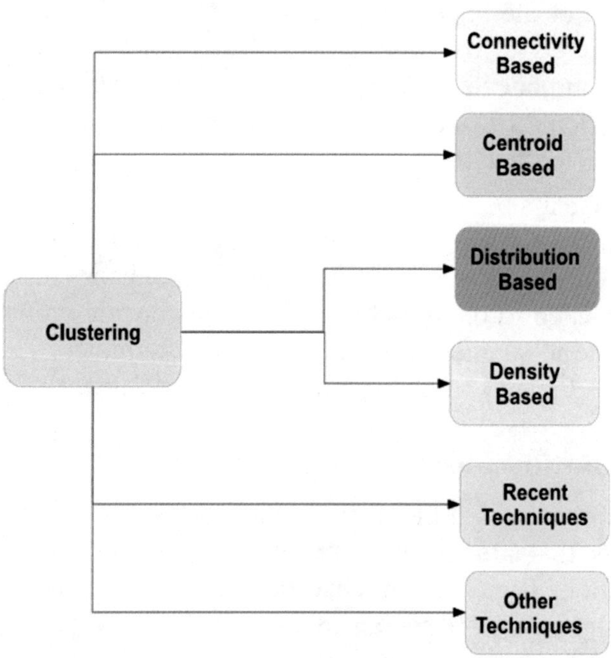

Figure 2.3. Types of Clustering and the x-axis, for placing the objects, ensuring that the clusters do not mix.

Source: https://www.researchgate.net/figure/Types-of-Clustering-and-the-x-axis-for-placing-the-objects-ensuring-that-the-clusters_fig2_330213114

There are several variants of Connection-based clustering depending on how the lengths are estimated, such as Unweighted Pair Group Method with Arithmetic Mean (UPGMA), also known as Average Linkage Clustering, Single-Linkage Clustering, which includes calculating the shortest distance between objects, and Complete-Linkage Clustering, which computes the highest of object distances (Backes et al., 2007). Divisive or Agglomerative Clustering could be used to choose relevant clusters from the given hierarchy of clusters. We start with single items in Agglomerative Clustering and conglomerate these into clusters, but in Divisive Clustering, our start is with the entire data set and separate it into segments (Backes et al., 2007).

2.3.2. Centroid Based Clustering

Centroid-based clustering can have clusters that are described by a vector that isn't absolutely a data set part, or clusters that are tightly limited to the dataset's components. The clusters quantity in the kmeans Clustering method is restricted to size k, hence determining centers of k cluster and allocating items to the closest centers is essential.

To determine the best of numerous runs, the method is performed based on various random k initializations. Clusters in kmedoid clustering are tightly limited to dataset members, whereas in kmedians clustering, just the medians are picked to build a cluster. The fundamental drawback of these methods is that the quantity of clusters k is predetermined. They also result in erroneously clipped borders among the clusters (Kumar et al., 2013).

2.3.3. Distribution Based Clustering

Clusters are formed using the distribution-based clustering method, which selects objects that are much more probable to correspond to the very same distribution. The Gaussian Distribution is among the most often used distribution algorithms. It has an overfitting issue, which occurs when a model cannot adapt to the training data set.

2.3.4. Density-Based Clustering

In this sort of clustering, a cluster is defined as an area with a greater density than that of the remaining data set. Border and Noise points are regarded to be sparse areas objects. OPTICS, Mean-Shi SCAN, and DBSCAN are based on linking sites that meet a density criterion while staying under defined distance limits. All density-connected items and objects inside these objects' ranges are free to get any shape within the cluster that is produced.

2.3.5. Recent Clustering Techniques

On high-dimensional information, most standard clustering approaches fail, hence new strategies are being investigated. Correlation and Subspace Clustering are two types of clustering algorithms. The clustering model in Subspace Clustering describes a shortlist of qualities that should be examined for the cluster development, whereas in Correlation, the model includes the correlation among the selected qualities in addition to the list of qualities (Braun et al.m 2020).

2.3.6. Other Techniques

The Basic Sequential Algorithmic Scheme (BSAS) is among the most fundamental clustering algorithms. Even though the quantity of clusters to be generated is unknown, the BSAS generates them using the length d(p, C) between a cluster C and a vector point p, the highest quantity of clusters permitted q and the limit of disparity 0.

Based on the route to the current clusters, each newly supplied vector is either allocated to a current cluster or a fresh cluster is generated.

2.3.7. Clustering applications in IDS

In the operation of intrusion detection, the clustering method can be quite useful. Figure 2.4 depicts the setup. Multiple IDSs, both Host and Network types produce alerts that are logged into a centralized system. The alarm messages received from various IDSs will be in various formats. A preprocessing phase is required before sending information to the server to ensure that they are all in the same format (Herman et al., 2012).

In the preprocessing phase, maximum endeavor values are selected for the missing properties. For comparability, the timestamp data may need to be transformed into seconds. Because various IDSs may be using various naming conventions for the same event, it is necessary to standardize the messages.

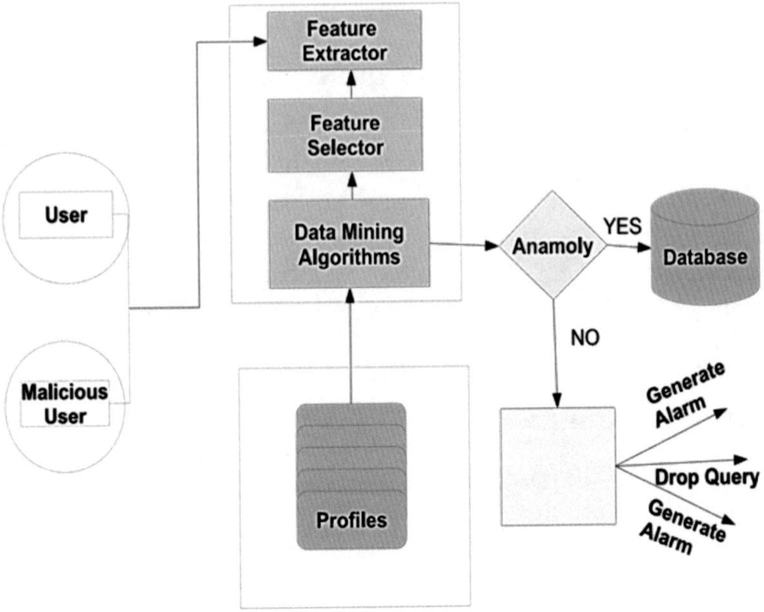

Figure 2.4. Use of Clustering in Intrusion Detection System (IDS)

Source: https://globaljournals.org/item/6640-security-in-data-mining-a-comprehensive-survey

To maintain a record of the alerts, each one might be given a different ID. These are sent to the first step for filtration and labeling functions after being preprocessed and normalized. To reduce the number of Alerts, it's a great idea of utilising Alert Fusion, which combines alerts with similar features that differ by a tiny period. The generalization process is sped up with Alert Fusion. The process of generalization entails incorporating hierarchical prior knowledge into every feature. The specified property is generalized towards the subsequent highest degree of hierarchy on each iteration of this operation, or those alerts that have grown comparable by now are grouped (Braun et al., 2020).

2.4. PRIVACY-PRESERVING DATA MINING (PPDM)

The goal of Privacy Protecting data mining methods is to extract important information from enormous amounts of data while safeguarding any sensitive data that may be present. It protects sensitive data to safeguard privacy while yet enabling us to complete all data mining processes quickly. The two sorts of data mining methods that are concerned with privacy are:

- Confidentiality of data
- Confidentiality of information

Information privacy concentrates on the alteration of the data for the security of critical information that may be extracted from the data, whereas data privacy concentrates on the alteration of the dataset for the security of critical information of users.

Information privacy, on the opposite hand, is concerned with giving security to the input, whilst Data privacy is focused on maintaining protection to the output. A PPDM system's principal goal is to keep confidential info hidden from prying eyes (Shapiro, 2012). The PPDM methods rely on examining mining methods for any adverse effects obtained during information privacy. The goal of Privacy-Preserving Data Mining is to create methods that change the original information in such a way that all private knowledge and data remain hidden well when the mining process is completed.

Occasionally, various parties may seek to collaborate. In principle, data mining technology is private data arising from effective aggregation and privacy-neutral. The purpose in which a data mining method is employed without releasing any sensitive data may be good or bad. Consider the harmful nature of many book stores. Data mining has broadened the with corresponding sales data which is in some ways regarded to inquiry possibilities, allowing scientists to be extremely sensitive, could desire to share partial manipulate enormous sets of data on the one hand, whereas the data among themselves to start arriving at the malignant usages of these methods on the other hand agglomeration patterns without divulging their participant s For data protection, this necessitates the usage of secure protocols (Pinkas, 2002). Distributing information to a large number of people Privacy In such circumstances, determining the foundation of privacy-preserving data must be accomplished with high degrees of mining accuracy (Xia et al., 2020).

The necessity of the hour is for algorithms and associated privacy technologies. We must address a few queries in this concern, including:

- Assessment of various algorithms concerning one another
- Would each one of the data mining methods use privacy-preserving techniques? Or is it for all of them?
- Increasing the areas where these strategies can be used.

- Examining their uses in defense and geospatial, intelligence, inspection applications.
- Data mining strategies that combine trust, confidentiality, and privacy as well as strong opinions.

Table 2.1. Research Progress in PPDM

Authors	Algorithm	Performance	Future enhancement
Boutet et al.(2012)	kNN	Better than Randomization scheme	Can consider all attacking models
Tian et al.(2015)	Correlated Differential Privacy (CDP)	Enhances the utility while answering a large group of queries on correlated datasets	Can be experimented with Complex Applications
Bharath et al.(2015)	PP k-NN classifier	Irrespective of the values of k, it is observed that SRkNNo is around 33% faster than SRkNN. E.g., when k=10, the computation costs of SRkNNo and SRkNN are 84.47 and 127.72 minutes, respectively (boosting the online running time of Stage 1 by 33.86%)	Parallelization is not used
Nethravathi et al.(2015)	PPDM	Reduced misplacement clustering error and removal of data that is sensitive and correlated	Works only for numerical data
Mohammed et al.(2013)	Differential Privacy	More secured under the Semi-Honest model	Overcoming Privacy Attack
Vaidya et al.(2013)	Distributed RDT	Lower Computation and Communication cost	The limited information that is still re- value must be checked
Lee (2014)	)Perturbation methods	Capable of performing RFM Analysis	Partial disclosure is still possible

2.4.1. Distributed Privacy-Preserving Data Mining(DPPDM)

The explosive rise of the internet in recent years has opened up new possibilities for distributive data mining, wherein mining processes are carried out collaboratively utilizing private inputs Mining operations among untrustworthy parties or rivals frequently lead to information leakage (Friedman & Schuster, 2010). As a corollary, Distributed Privacy-Preserving Data Mining (DPPDM) techniques necessitate a significant degree of cooperation among parties to derive conclusions or exchange non-sensitive mining findings. This could lead to the revealing of critical data in some cases.

Vertically segmented data and Horizontally segmented data are two types of distributed data mining. Every site in a horizontally segmented data architecture keeps comprehensive data on a distinct set of entities, and the integrative database is the sum of all these databases. On either hand, in a vertically segmented data structure, every site maintains various sorts of data and each database has only restricted data about the same set of things. The data leakage produced by distributed computational approaches can be limited using privacy features (Roughan & Zhang, 2006).

Every non-trusting participant can calculate its very own functions for specific inputs set while disclosing the functions' defined outputs. The confidentiality service not only hides critical information, but it also manages it and its utilization through a series of agreements and tradeoffs among concealing and revealing. All effective PPDM methods are predicated on the idea that releasing intermediary obtained results during data mining processes is appropriate. Encryption algorithms resolve the issue of privacy protection, and their use could enable data mining activities between mutually untrustworthy individuals, as well as between opponents, much easier. Distributed Data Mining Techniques use encryption methods to protect data privacy. Cryptography is employed in both methods to Distributed Data Mining (vertically and horizontally segmented data) without putting much emphasis on the efficiency of the encryption mechanism utilized (Mohammed et al., 2013).

Horizontal segmentation occurs when information is saved on various machines and segmentation is done row-by-row, while vertical segmentation occurs when information is saved and segmented column-by-column. Fig.2.5 depicts a high-level picture of the situation.

Instead of divulging data on individuals, the goal of data mining methods is to develop higher-level principles or descriptions and generalize across populations. It works by analyzing personal information that is sensitive to privacy issues. Because much of this data is currently in the hands of many entities, maintaining privacy is a major challenge (Kumar et al., 2019).

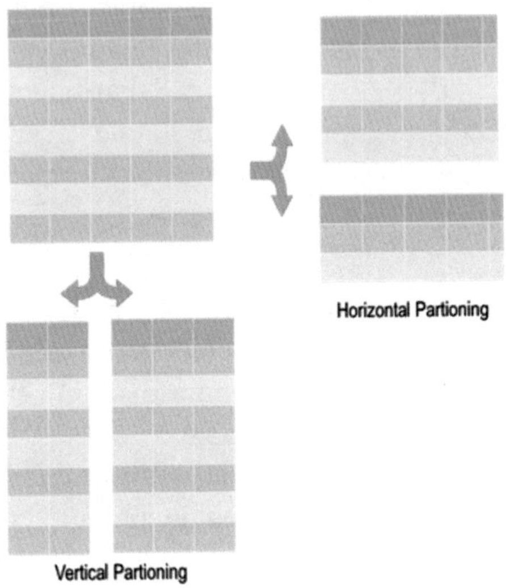

Figure 2.5. Horizontal and Vertical Partitioning Techniques

Source: https://globaljournals.org/GJCST_Volume16/6-Security-in-Data-Mining.pdf

Individual and control safeguards should be segregated to offer appropriate security and prevent any potential linking of this data. Regrettably, this division makes it harder to utilize the data for reasons such as detecting illegal conduct and other socially beneficial objectives. Plans to exchange data between agencies to fight terrorism, as well as other illegal activities, would also abolish the segregation barriers (Vaidya et al., 2013).

A risk model which concentrates on the implementation of FIPPs (Fair Information Practice Principles) is insufficient in so numerous complex socio-technical systems.

2.4.2. Obfuscation of Public-Key Program

Program obfuscation is the method of making a program unintelligible without affecting its functionality. An obfuscated program should be a virtual black-box, which means that if it is feasible to calculate something out of it, it should also be necessary to calculate the right thing from the program's input-output behavior. Obfuscation of public programs can be categorized as Single-Database Personal Information Retrieval (Lee, 2014). A public-key program obfuscation role consolidates p into (*Dec*, *P*), where *P* on any input calculates cryptography of what *P* would calculate on the identical input, and the decryption method *Dec* decode the result of P, provided a program *P* from a category of programs *C* and a safety variable *s*. Which is, *Dec* (*p(i)*) = *p(i)* for any input *i* but it is infeasible to tell which p make the category C was used to generate P for a provided code P. The span of the program encryption |P| must be polynomially dependent on only s and |*p(i)*|, and the span of the result |*p(i)*| must be polynomially dependent on the only *k* and |*p(i)*|.

2.4.3. Multi-party Computation in a Secure Environment

Distributed computer technology entails the use of multiple, interconnected data processing gadgets to perform a consolidated computation of a feature. Servers running a distributed collection of the data system, for instance, may want to keep updating their collection of data. Reliable multi-party computation's goal is to enable parties to perform distributed computing duties in a secure manner (Backes et al., 2007). It usually entails the parties performing a computation depending on their inputs, with none of them ready to share one's input with others. The challenge is to carry out this kind of computation while maintaining the confidentiality of their input data. The SMC - Secure Multi-party Computation - the issue is the name given to this problem (Braun et al., 2020). Consider the situation where two parties want to calculate the median securely. Both parties have two input sets, X and Y, with them. The parties must calculate the median of the union of their sets X U Y without disclosing anything regarding the of others set. In a surrounding where various data holders have various kinds of info regarding a widely known set of elements, Association Rules can be calculated.

2.5. INTRUSION DETECTION SYSTEM (IDSIDS)

Intrusion detection systems are designed to identify an intrusion as soon as it occurs (Chourse & Richhariya, 2008). The innovation of a detailed IDS requires a higher level of human competence and a substantial amount of time .

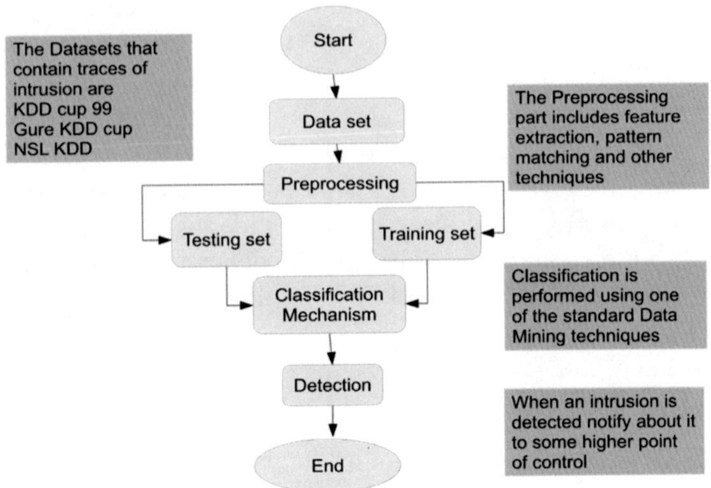

Figure 2.6. An Intrusion Detection System Overview

Source: https://www.semanticscholar.org/paper/Security-in-Data-Mining-A-Comprehensive-Survey-Niranjan-Nitish/0ab17a70020d4e617325fbb3855fe002c9264547

Rules, on the other hand, An intrusion detection system can be IDSs relying on DMT - Data Mining techniques - that are categorized depending on the category of intrusion and location and involve less expertise while performing better. Sensors identify web threats and send alerts to a core engine relying on the strategy utilized by the Intrusion Detection System (Rajasekaran & Nirmala, 2015). The majority of threats against susceptible services, including IDS installations, include all privilege escalation, components integration, and three data-driven on applications into a single device. Unregistered logins and connections directly to files are susceptible, according to the latest virus scanner methodology (Jabez & Muthukumar, 2014). A Detector depending on signatures senses viruses from code efficiently, and a Classification depending on heuristic rules for sensing can be utilized as a cybersecurity tool. A look at the latest malware. The signature-based tracking of an IDS - Intrusion Detection System - is depicted in Figure 2.6. The algorithm depends on signatures, which are distinctive cords, to generate elements like detection models, sensors, and a console monitor. The drawbacks of this strategy include the central engine (Cárdenas et al., 2014). Security happenings generated by sensors are much more time-consuming and fail to identify new threats, whereas all alerts and activities are controlled and monitored malevolent executable code.

On the other hand, Heuristic classifiers, are produced by a group of virus specialists for happenings in a dataset and produce warnings relying on a set identification of new malevolent executables, according to the Central Engine records and the Console Monitor.

2.5.1. Kinds of IDS

Intrusions can be spotted on a single computer or across a network, and as a result, intrusion protection technologies are widely utilized as the first line of defense. Three kinds of Host Mechanisms, IDS, and namely Network-Based, are not appropriate to counter attacks as a sole defense. Hybrid and Based IDS are two types of IDS.

2.5.1.1. IDS Based on Network

Computer networks have been attacked by criminals and enemies due to their growing importance in contemporary societies. Finding the best feasible alternatives is critical for the security of our systems (Lu & Traore, 2004). As the first step of defense, intrusion prevention strategies like biometric/password authentication, information protection using encoding, and programming error avoidance techniques have all been extensively used. Intrusion prevention strategies are insufficient as a sole defense system against attacks. As a result, it can only be utilized as a second step of defense for computer network security (Burlakov et al., 2016).

An IDS - Intrusion Detection system - must safeguard resources like the kernels of a target system, user accounts and file systems, and must be allowed to characterize the normal/valid behavior of these resources using strategies that analyze continuing network activities with previously developed designs and define intrusive activities (Balasingam et al., 2017). The source of data for Network-dependent Intrusion Detection Systems is network packets. A network adapter is used by the NIDS to analyze and pay attention to network activity as packets move all across the system. When an incursion from out of the enterprise's perimeter is detected, a network-based IDS produces alerts. To examine both outbound and inbound packets, network-based IDSs are unequivocally positioned at tactical locations on the LAN. Inbound packets that may override the firewall are detected by network-based IDSs, which are installed beside the firewalls (Abraham et al., 2007). Few Network-Based IDSs accept specially made signatures from the customer's security policy as input, allowing for only confined identification of policy violation. The IDS may not be capable of detecting intrusion packets that arise from authorized users (Li & Guo, 2007).

The following are a few of the benefits of a network-based IDS:

- Have the ability to supervise larger networks.
- They can be made inaccessible for added protection against attacks.
- Integrating an IDS into an operating network is simple.
- They can operate without tampering with a network's routine operation (Mitchell & Chen, 2013).

The following are the drawbacks:

- Their success is frequently determined by the functionalities of the network's transitional switches.
- If the assailants fragment and discharge their packets, the IDS may become crash and unstable.
- Incapable of decrypting encrypted information sent over a virtual private network (VPN).

2.5.1.2. IDS Based on Host

The surveillance detectors in a Host-based IDS are installed on system resource endpoints to observe logs produced by application programs or the Host Operating System.

These Audit log files keep track of activities and events at individual Network resources (Yeung & Ding, 2003). Because a host-based IDS can identify assaults that a system-based IDS cannot, a hacker can take advantage of one of your trustable insiders. A Host-based method employs a Signature Rule Base that is inferred from site-specific security policies. Because a Host-Based IDS can alert security guards with the site specifics of an invasion, he can take swift action to prevent the incursion, it can surmount all of the issues affiliated with a Network-Based IDS. A host-based IDS also can keep track of an attacker's failed endeavors. It can also keep separate documentation of client login and logoff activities so that audit databases can be generated.

The following are some of the benefits of using a host-based IDS:

- Uses audit log paths from operating systems to identify threats entailing software credibility infringements

- Can identify attacks that a network-based IDS are unable to detect.

The following are the drawbacks:

- It is not designed to detect network-based attacks.
- Each independent system is complicated to manage and configure
- They can be disabled by specific kinds of DoS (Denial of Service) attacks (Tseng et al., 2003).

2.5.1.3. Hybrid IDS

Because host and network-based IDSs have advantages and strengths that are distinct from one another, combining the two methods into next-generation IDSs is a great idea (Vittapu et al., 2015). A Hybrid IDS is a term used to describe this kind of combination. The insertion of these two elements would significantly improve resistance to a few additional assaults.

2.5.1.4. DM Methods for IDS

The following are some of the data mining methods and applications required by IDS:

- Classification
- Feature Selection and
- Pattern Matching

2.5.2. Pattern Matching

Pattern matching is the method of obtaining required data by locating a specific pattern of a portion of information - binary or substring pattern - in the entire packet (Dubiner et al., 1994). It is straightforward to use, despite its rigidity. Only when the data in query is connected with a specific service or is intended for or from a specific port does a Network-Based IDS excel in identifying an attack. That is, only some fields of the data must be inspected, like the Destination/Source port address, the Service, and some others, decreasing the quantity of packet investigation required. It does, after all, make it more complicated for mechanisms to cope with Trojans and their affiliated traffic, which can be shifted at any time. Depending on the intensity of onset, pattern matching can be divided into two groups:

- Pattern Matching with Frequent Patterns and
- Pattern Matching with Outlier Patterns

2.5.2.1. Frequent Pattern Matching

These are the trends that appear widely in audit information, i.e., their intensity of occurring is higher than that of other trends in the identical data (Soni & Sharma, 2014).

Identifying common trends in big data aids in the analysis and prediction of specific data qualities. Frequent pattern matching, for instance, could aid in predicting future sales outcomes by analyzing an agency's sales data. It also aids in the making of decisions.

Mining the repository for incursion-free (train) information and comparing it to the trends of the ordinary portfolio - train - data is a common pattern mining technique used in ADAM project information. To decrease false-positive aspects, a classifier is being utilized.

They are uncommon and take place infrequently, so they will have little assistance in the records. These trends can often indicate a data disparity, like economic recession, fraudulent transactions, abnormal behavior, intrusion, and so on. Outlier pattern mining techniques can be of two kinds: one that calculates and monitors trends all of the time, and the other that makes it look for designs only at predetermined time intervals. Outlier papers employ data structures like String Matching Algorithms and the Suffix Tree to find outliers. The dataset's dimensional space We use Feature Selection to accomplish this.

2.5.2.2. Feature Selection(FS)

Feature Selection is the method of decreasing the set of data aspects by choosing a subset of the characteristics from a provided set of attributes (Vasanthakumar et al., 2017). FS entails removing features that are irrelevant and redundant. FS is a machine learning method that aids in the development of effective classification processes. The time intricacy of a classifier is decreased with enhanced precision as subset dimensionality is decreased. Information Gain is an attribute selection hypothesis that can be utilized to calculate the entropy cost of each characteristic. A position can be used to describe an entropy cost. Each feature's significance or affiliation with a solution group that is employed to recognize the info is represented by its rank. As a result, a character with a higher rating will be one of the most essential categorization attributes. Filter technique, wrapper technique, and embedded technique are the three standard strategies for feature selection that are frequently used.

2.5.2.3. Outlier Pattern Matching

Classification: For studying a framework and classifying samples the existing trends and which are not sound are of information into predefined classes, the classification uses different and unusual from examples training patterns. Outlier Patterns cover a broad spectrum of phenomena. Sound is eliminated using preprocessing phase categorization methods such as Neural Networks because it is not a portion of the real data, Bayesian Belief, Bayesian Classifier, Decision Trees (Farid et al., 2014).

On the other hand, outliers, are impossible to eradicate. Outliers display departing attributes when compared to Data Mining algorithms, so systems and others are employed in applications. The majority of other scenarios are usually classified.

Table 2.2. IDS Research Progress

Algorithm	**Performance**	**Future enhance-ment**	**Authors**
Hyperboli Hopfiel Neural Network(HHNN)	90% Detection rate	Can be enhanced	Jabez J et al. (2014)
Behavior Rule Analysis	good perfor-mance	Can be analyzed with other methods	Mitchell et al. (2013)
Feature Selection	good classifica-tion	NSL-KDD Can consider	Soni et al. (2014)
SVM Classification	FPR of 5% and TPR of 96%	Other techniques can be used as well.	M Vittapu et al. (2015)
Genetic Fuzzy System	In terms of the mean F-measure, average accu-racy, and false alarm rate, this is the best trad-eoff.	It's possible to use a Multi-objective Evolutionary Al-gorithm to maxi-mize performance metrics.	S Abadeh et al. (2007)

Outliers patterns do not necessitate the following steps:

- The first step is to create a training dataset.
- Determination of attributes and classes
- Identifying attributes that can be used to classify things

- Analyze the relevance
- Using training examples to learn the model
- Practice with the set
- Using the model to classify unidentified data samples

2.6. CLASSIFICATION OF PHISHING WEBSITES

Phishing is the practice of imitating a legitimate and well-known company's website to steal users' personal information (password and username). Deceitful people frequently make artificial websites to imitate legitimate websites. Users unwittingly lose cash as a result of hackers' phishing schemes. As a result, online buying and selling necessitate safety from such attacks, which is regarded as a crucial step. The quality of the derived features determines the precision of a website's classification and prediction. Most web users believe that using an antimalware tool protects them from phishing attacks, so antimalware tools must be precise in anticipating phishing. Phishing websites leave a trail of clues in their content and via browser security indicators. To combat the issue of phishing, several options have been suggested. Data mining methods that use Rule-based classification have shown to be effective in predicting phishing assaults (Marchal et al., 2014).

Usually, a phishing threat begins with an attacker who sends an email to victims asking for private information by accessing a specific URL. To bring out appropriate deceit, phishers use a collection of mutual attributes to develop phishing websites (Kumaraguru et al., 2010). Based on the derived characteristics of the webpage visited, we can use this data to effectively differentiate between phishy and non-phishy websites. The two most common methods for detecting phishing webpages are black-list centered, which compares the suggested URL to those on the list, and heuristic-based, which collects specific characteristics from the webpage to classify it as legitimate and phishy. Because a new malevolent website is initiated each second, a black-list-based strategy has the drawback of not being able to restrain all phishing sites. A Heuristic-based method, on the other hand, can detect new frauds websites (Fu et al., 2006). The achievement of heuristic-based strategies is determined by the attributes chosen and how they are analyzed. Data mining can be utilized to identify patterns as well as relationships between them. Decisions are made relying on rules and patterns obtained using data mining techniques, so data mining is deemed significant for making decisions (Chen & Guo, 2006).

2.7. ARTIFICIAL NEURAL NETWORKS (ANN)

An Artificial Neural Network (ANN) is a collection of interconnected processing units. Each linkage has a weight assigned to it that defines how one unit impacts the otherSome Some of these units serve as input points, while the rest serve as output points. The secret layer is made up of the remaining points. By triggering each input point and enabling it to disperse via the secret layer points to the output endpoints, a neural network conducts a functional mapping from input data to output data. The navigation is saved as a weighted average of the connections.

The structure of HHNN is depicted in Figure 2.7 (Jabez & Muthukumar, 2014).

In the domain of intrusion detection, ANN is among the most frequently used method. Three types of artificial neural networks (ANNs) exist:

ANN is one of the widely used techniques in the field of intrusion detection. ANN techniques are classified into three categories namely:

- Unsupervised Intrusion Detection,
- Hybrid Intrusion detection, and
- Supervised Intrusion Detection

Multi-Layer Feed Forward (MLFF) Recurrent Neural Networks and Neural Networks are examples of supervised intrusion detection relying on ANN. The Neural networks of MLFF can readily achieve the local minimum since the amount of training data sets is large and its dispersion is unbalanced, leading to poorer stability. With fewer regular assaults, the accurate rate of an MLFF neural network is poor. The identification accuracy of supervised IDS is worse than that of Multivariate and SVM Adaptive Regression Splines (MARS). Unmonitored Intrusion Detection relying on ANN classifies testing data and distinguishes between abnormal and normal actions (Fakoor et al., 2020). This can substantially enhance the research on new data because it does not require retraining. For low-frequency attacks, the efficiency of Unmonitored ANN is also poorer, resulting in reduced detection precision. To address the constraints of the basic types of ANN, hybrid approaches merging unsupervised and supervised ANN, as well as integrating ANN with the other data mining methods for intrusion detection, could be accomplished. In comparison to Intrusion Detection relying on RBF networks only, a hybrid technique integrating RBF and SOM networks is significantly more effective. A hybrid system that combines Particle Swarm Optimization (PSO), Flexible Neural Trees, and Evolutionary Algorithms is an extremely effective hybrid ANN, which combines ANN with the Fuzzy Clustering technique, splits the training set into sub - sets, boosting the stability of each ANN for low-frequency attacks (Chen & Guo, 2006). As a result, one can claim that hybrid ANNs has been the tendency in intrusion detection relying on ANNs. The effectiveness of intrusion detection is influenced by various techniques of generating Hybrid ANN. As a result, multiple Hybrid ANN systems must be built to fulfill distinct aims. The HHNN(Hyperbolic Hopfield Neural Network) is one of the Hybrid techniques used for intrusion detection. Anomaly detection believes that incursions always manifest themselves as a series of departures from the norm. The HHNN method examines the link among the two data sets and generalizes it to generate additional input-output pairings that are acceptable. In theory, neural networks might be utilized to identify assaults and search for them in the auditing stream. Because there is currently no credible method for identifying reasons for linkage, it is impossible to determine the rationale for the attack's categorization. Table 2.3 summarizes the existing progress made in HHNN (Bui et al., 2016).

2.8. OUTLIER DETECTION/ANOMALY DETECTION

Anomaly detection is the method of identifying trends that do not match the expected behavior. Anomalies are designs like this. Contaminants, outliers, peculiarities, surprises, and aberrations are all terms used by different application domains to describe them. The terms "outliers" and "anomalies" are frequently used in this perspective. Detection of fraud for debit and credit cards, insurance, and health care are examples of anomaly detection application domains. It is also used to detect intrusions and faults in a security system, as well as enemy activity detection (Song et al., 2007).

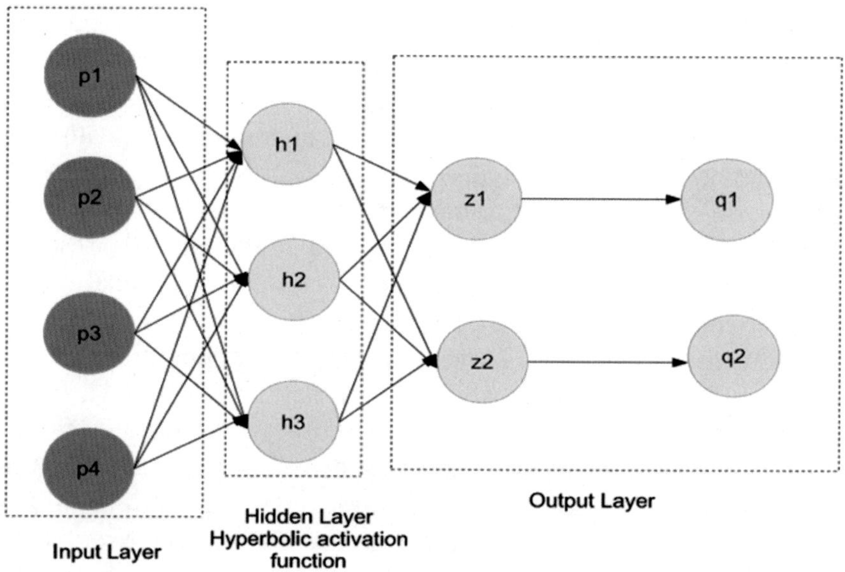

Figure 2.7. Hopfield Neural Network (HHNN) overview

Source: https://www.tutorialspoint.com/artificial_neural_network/artificial_neural_network_hopfield.htm

Table 2.3. ANN Research Progress

Algorithm	**Performance**	**Future enhancement**	**Authors**
Kappa and ROC Index	SVM (88.7 %), MLP (90.2 %), KLR (87.9 %), LMT (86.1 %) and RBF (87.1 %)	It is possible to use the Information Gain Ratio as a feature selection.	D T Bui et al. (2015)
Artificial neural network analysis and learning theoretical framework	Improves the performance of generalization.	Various optimization techniques and network designs can be used.	C Corteset al. (2016)

The majority of abnormal trends distract from the norm. Anomalies are plotted on two-dimensional information set in Figure 2.8. The zones N1 and N2 are regarded as normal though since they contain a large proportion of the observational data. Anomalies are points O1 and O2 which are much further ahead of such regions, as well as points in region O3. Anomalies are introduced into data for a variety of causes, but they all share the trait of being intriguing to investigate. Outliers' fascination is a key attribute of anomaly and outlier identification. Anomaly detection is similar to, but not identical to, noise accommodation, and removal of noise, in that it must cope with undesired noise in the records (Hwang et al., 2003). For the researcher, noise is an unwelcome component that obstructs statistical analysis. As a result, noise removal is required because undesired statistics must be taken away before data evaluation. Novelty Sensing is a subtopic of anomaly detection that looks for previously unknown new trends in data. It differs from Anomaly Detection in which the sensed novel trends are embedded into the standard model. Various anomaly detection alternatives will also operate for unique detection, and vise - versa. As a result, in Anomaly Detection, an area is outlined, with findings that comply to the region being deemed normal and non-conforming findings being regarded as anamolous (Liu et al., 2015).

2.8.1. Challenges

The following are some of the difficulties that scientists face when it comes to anomaly detection (Liu et al., 2015; Ni et al., 2016):

- It's hard to define a normal region in which all normal behaviors exist. The difference between abnormal and normal behavior is very narrow, so an assessment near the demarcation line could be normal, and vise - versa.
- When the intruders disguise themselves to start making the anomalous findings seem very normal. Defining
- Normal behavior becomes perplexing.
- Normal behaviors change over time, and what is regarded normal today may not be so in the future.
- Anomalies are defined differently in various application areas. In the healthcare domain, variations in body temperature are considered abnormal, whereas variation in the marketing domain may be regarded as normal. As a result, the application of a strategy that has been formed cannot be generic. Anomaly detection technologies involve labeled information for coaching and validation of designs, which is not easily available. It's difficult to separate noise from statistics and eliminate it. As a

result, dealing with the anomaly detection problem is difficult. The most anomaly detection method can only overcome a domain-specific issue formulation that is influenced by variables like anomaly types, data category, availability, labeled data, and so on.

2.8.2. Data Mining Mechanisms for Detection Anomalies

The following two strategies can be used to enforce an intrusion detection process (Wu et al., 2014):

- Signature Based IDS and
- Anomaly Based IDS.

Signature-based IDS detects invasions by blacklisting assault signatures that have been specified.

It is vulnerable to evasion processes because it is inefficient toward new kinds of assaults.

On the contrary, Anomaly-based IDS records ordinary behavior and categorizes divergences from it as anomalies. It is thought to be resistant to unidentified attacks and to protect against attacks by malevolent clients who enhance their assault technique. Anomaly Based IDS is extensively used because it makes substantial use of information.

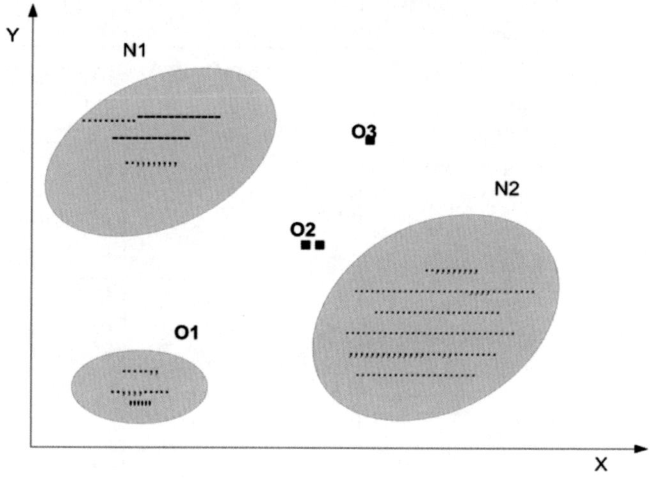

Figure 2.8. Outlier Detection

Source: https://globaljournals.org/GJCST_Volume16/6-Security-in-Data-Mining.pdf

Table 2.4 Code Injection Attacks Research Progress

Algorithm	Performance	Future enhancement	Authors
Input based approach		More SQLi avoidance techniques/tools can be researched or developed.	G Parmar et al.(2015) (118)
Contextual Fingerprinting	On the execution time, there is an 11.1 percent overhead.	It is possible to lower the overhead.	Mitropoulos et al.(2016) (116)
Exception Oriented Programming (EOP)	A 90% detection rate is expected.	It's possible to extend this to Mac and Windows kernels.	L Deng et al.(2016) (119)
Dynamic Binary Instrumentation	There is a 2.4x overhead, which is comparable to other methods, but there are no false alarms.	It is possible to lower the overhead.	A Follner et al.(2016) (117)
Emulation-based framework for ROP	With a 16-core Intel E5-2630 (2.3GHz) processor and 24GB RAM, the study took 4 hours.	Total analysis time is cut in half.	M Graziano et al.(2016) (115)
Cross-Site Scripting Secure Web Application Framework	Depending on the type of JSP program, the percentage can range from 1.25 percent to 5.75 percent.	Discovering strategies for removing the HTTP response delay and other XSS-SAFE rule checks without affecting the detection efficiency of XSS attacks.	S Gupta et al. (2016) (120)

Two phases are involved in mining techniques:

- The Phase of Detection
- The Phase of Training

Profiles are formed during the training period by categorizing normal access behaviors and forwarded to the Feature Selector, classifier, and Feature Extractor in batch mode. Out of standard access behavior, the Classifier creates a trained model.

2.8.2.1. Code Reuse Attacks

Code Reuse Attacks are invasions where an attacker redirects control flow via previously written code, resulting in an incorrect result (Habibi et al., 2015).

As a result, attackers have developed code-reuse attacks, where a software flaw is manipulated to generate a control flow via an established source code to a malevolent end. Return Into Lib C (RILC) is a form of code-reuse attack in which the heap is breached and regulation is transmitted to the start of an established library function like mprotect() to generate a memory region that enables both execution and write operations to override W+X. Data Mining algorithms can be used to effectively counteract such attacks. If there are any kind of flaws in the reference code, the guidelines are categorized as malevolant (Mohanappriya & Vijayalakshmi, 2018). SVM, Bayesian, and Decision Tree are some of the classifying techniques that can be utilized in this respect (Checkoway et al., 2010).

2.8.2.2. Return Oriented Programming

ROP assaults begin when an assailant acquires stack control and diverts it to a tiny snippet of script known as a gadget, which usually ends with RET guidelines. Since attackers influence return addresses, they can allocate the RET of one device to the beginning of another, getting the required functionality from a small collection of small devices (Bugiel et al., 2102). Although ROP attacks do not inject code, they can cause arbitrary behavior in the targeted system. To counter any type of ROP, a compiler-based method has been proposed in . The writers of current in-place code randomization, which can be used to reduce ROP assaults on third-party software. Return-oriented vulnerabilities are feasible to write, according to Buchanan et al. (2008), because the intricacy of device combinations is abstracted behind a coding compiler and language. Davi et al. (2009) suggested run-time integrity monitoring methods depend on dynamic tracing and taint assessment, which involves tracking orchestration of program binaries. A tool called DROP is described in, which dynamically identifies ROP malevolent code.

2.8.2.3. Jump Oriented Programming

An attacker uses a limited collection of indirect JMP guidelines rather than RET guidelines to connect the devices in JOP - Jump Oriented Programming. Among the devices, a special device known as a dis-patcher is utilized for flow control management (Yao et al., 2013).

REFERENCES

1. Abadeh, M. S., Habibi, J., & Lucas, C. (2007). Intrusion detection using a fuzzy genetics-based learning algorithm. *Journal of Network and Computer Applications*, *30*(1), 414-428.
2. Abraham, A., & Jain, R. (2005). 1 Department of Computer Science, Oklahoma State University, USA aj ith. abrahamSieee. org 2School of Information Science, University of South Australia, Australia ravi. j ainQunisa. edu. au. *Classification and Clustering for Knowledge Discovery*, *4*, 191.
3. Abraham, A., Grosan, C., & Martin-Vide, C. (2007). Evolutionary design of intrusion detection programs. *Int. J. Netw. Secur.*, *4*(3), 328-339.
4. Abraham, A., Jain, R., Thomas, J., & Han, S. Y. (2007). D-SCIDS: Distributed soft computing intrusion detection system. Journal of Network and Computer Applications, 30(1), 81-98.
5. Aggarwal, C. C. (2005). On k-anonymity and the curse of dimensionality. In *VLDB*, 5, 901-909.
6. Ali, A., Akhtar, N., Khan, B. A., Khan, M. S., Rasul, A., Khalid, N., ... & Ali, L. (2012). Acacia nilotica: a plant of multipurpose medicinal uses. *Journal of medicinal plants research*, *6*(9), 1492-1496.
7. Backes, M., Pfitzmann, B., & Waidner, M. (2007). The reactive simulatability (RSIM) framework for asynchronous systems. *Information and Computation*, *205*(12), 1685-1720.
8. Balasingam, B., Mannaru, P., Sidoti, D., Pattipati, K., & Willett, P. (2017). Online anomaly detection in big data: The first line of defense against intruders. In *Data Science and Big Data: An Environment of Computational Intelligence,* 2(1), 83-107.
9. Barrantes, E. G., Ackley, D. H., Forrest, S., & Stefanović, D. (2005). Randomized instruction set emulation. *ACM Transactions on Information and System Security (TISSEC)*, *8*(1), 3-40.
10. Beaver, D., Micali, S., & Rogaway, P. (1990, April). The round complexity of secure protocols. In *Proceedings of the twenty-second annual ACM symposium on Theory of computing*, 3(1), 503-513.
11. Bharath, G., Madhu, R., Chen, S. M., Veeramani, V., Mangalaraj, D., & Ponpandian, N. J. J. O. M. C. A. (2015). Solvent-free mechanochemical synthesis of graphene oxide and Fe 3 O 4–reduced graphene oxide

nanocomposites for sensitive detection of nitrite. *Journal of Materials Chemistry A*, *3*(30), 15529-15539.

12. Bhatkar, S., DuVarney, D. C., & Sekar, R. (2003, August). Address Obfuscation: An Efficient Approach to Combat a Broad Range of Memory Error Exploits. In *USENIX Security symposium*, 12(2), 291-301.
13. Bhatkar, S., DuVarney, D. C., & Sekar, R. (2005, August). Efficient Techniques for Comprehensive Protection from Memory Error Exploits. In *USENIX Security Symposium*, 10, 1251398-1251415.
14. Boutet, S., Lomb, L., Williams, G. J., Barends, T. R., Aquila, A., Doak, R. B., ... & Schlichting, I. (2012). High-resolution protein structure determination by serial femtosecond crystallography. *Science*, *337*(6092), 362-364.
15. Braun, L., Demmler, D., Schneider, T., & Tkachenko, O. (2020). MOTION-A Framework for Mixed-Protocol Multi-Party Computation. *IACR Cryptol. ePrint Arch.*, *2020*, 1137.
16. Buchanan, D., & Huczynski, A. (2004). Images of influence: 12 angry men and thirteen days. *Journal of Management Inquiry*, *13*(4), 312-323.
17. Buchanan, E., Roemer, R., Shacham, H., & Savage, S. (2008, October). When good instructions go bad: Generalizing return-oriented programming to RISC. In *Proceedings of the 15th ACM conference on Computer and communications security*, 2(1), 27-38.
18. Bugiel, S., Davi, L., Dmitrienko, A., Fischer, T., Sadeghi, A. R., & Shastry, B. (2012). Towards Taming Privilege-Escalation Attacks on Android. In *NDSS*, 17, 19.
19. Bui, D. T., Tuan, T. A., Klempe, H., Pradhan, B., & Revhaug, I. (2016). Spatial prediction models for shallow landslide hazards: a comparative assessment of the efficacy of support vector machines, artificial neural networks, kernel logistic regression, and logistic model tree. *Landslides*, *13*(2), 361-378.
20. Bui-Klimke, T. R., & Wu, F. (2015). Ochratoxin A and human health risk: A review of the evidence. *Critical reviews in food science and nutrition*, *55*(13), 1860-1869.
21. Burlakov, M. E., Golubyh, D. A., & Osipov, M. N. (2016). Naive bayesian classifier adaptation for e-mail classification mechanism. *Infokommunikacionnye tehnologii*, *14*(2), 199-203.

22. Cai, X., Ye, T., Liu, C., Lu, W., Lu, M., Zhang, J., ... & Cao, P. (2011). Luteolin induced G2 phase cell cycle arrest and apoptosis on non-small cell lung cancer cells. *Toxicology in Vitro*, *25*(7), 1385-1391.
23. Cárdenas, A. A., Berthier, R., Bobba, R. B., Huh, J. H., Jetcheva, J. G., Grochocki, D., & Sanders, W. H. (2014). A framework for evaluating intrusion detection architectures in advanced metering infrastructures. *IEEE Transactions on Smart Grid*, *5*(2), 906-915.
24. Checkoway, S., Davi, L., Dmitrienko, A., Sadeghi, A. R., Shacham, H., & Winandy, M. (2010). Return-oriented programming without returns. In *Proceedings of the 17th ACM conference on Computer and communications security*, 4(1), 559-572.
25. Chen, J., & Guo, C. (2006). Online detection and prevention of phishing attacks. In *2006 First International Conference on Communications and Networking in China*, 3(2), 1-7.
26. Chen, P., Xiao, H., Shen, X., Yin, X., Mao, B., & Xie, L. (2009, December). DROP: Detecting return-oriented programming malicious code. In *International Conference on Information Systems Security,* 9(12), 163-177.
27. Chen, Z., Wang, Y., Zhang, S., Zhong, H., & Chen, L. (2021). Differentially private user-based collaborative filtering recommendation based on K-means clustering. *Expert Systems with Applications*, *168*, 114366.
28. Cheng, K., Shi, J., Liu, Z., Jia, Y., Qin, Q., Zhang, H., ... & Liu, S. (2020). A panel of five plasma proteins for the early diagnosis of hepatitis B virus-related hepatocellular carcinoma in individuals at risk. *EBioMedicine*, *52*, 102638.
29. Chourse, S., & Richhariya, V. (2008). Survey Paper on Intrusion Detection Using Data Mining Techniques. *International Journal of Emerging Technology and Advanced Engineering, ISO*, *4*(8), 653-657.
30. Cortes, J. E., Saglio, G., Kantarjian, H. M., Baccarani, M., Mayer, J., Boqué, C., ... & Hochhaus, A. (2016). Final 5-year study results of DASISION: the dasatinib versus imatinib study in treatment-naïve chronic myeloid leukemia patients trial. *Journal of Clinical Oncology*, *34*(20), 2333.
31. Das, S., Chen, B., Chandramohan, M., Liu, Y., & Zhang, W. (2018). ROPSentry: Runtime defense against ROP attacks using hardware performance counters. *Computers & Security*, *73*, 374-388.

32. Dasgupta, D., Gonzalez, F., Yallapu, K., Gomez, J., & Yarramsettii, R. (2005). CIDS: An agent-based intrusion detection system. *Computers & Security*, *24*(5), 387-398.
33. Davì, G., & Patrono, C. (2007). Platelet activation and atherothrombosis. *New England Journal of Medicine*, *357*(24), 2482-2494.
34. Davi, L., Sadeghi, A. R., & Winandy, M. (2009). Dynamic integrity measurement and attestation: towards defense against return-oriented programming attacks. In *Proceedings of the 2009 ACM workshop on Scalable trusted computing*, 6 (1), 49-54.
35. Deng, L., & Zeng, Q. (2016). Exception-oriented programming: retrofitting code-reuse attacks to construct kernel malware. *IET Information Security*, *10*(6), 418-424.
36. Desale, K. S., & Ade, R. A Case Study: Stream Data Mining Classification. (2015). *International Journal of Computer Applications*, *975*, 8887.
37. Dubiner, M., Galil, Z., & Magen, E. (1994). Faster tree pattern matching. *Journal of the ACM (JACM)*, *41*(2), 205-213.
38. Dwork, C., McSherry, F., Nissim, K., & Smith, A. (2016). Calibrating noise to sensitivity in private data analysis. *Journal of Privacy and Confidentiality*, *7*(3), 17-51.
39. Fakoor, R., Mueller, J. W., Erickson, N., Chaudhari, P., & Smola, A. J. (2020). Fast, Accurate, and Simple Models for Tabular Data via Augmented Distillation. *Advances in Neural Information Processing Systems*, *33 (17), 1-18.*
40. Farid, D. M., Zhang, L., Rahman, C. M., Hossain, M. A., & Strachan, R. (2014). Hybrid decision tree and naïve Bayes classifiers for multi-class classification tasks. *Expert systems with applications*, *41*(4), 1937-1946.
41. Filková, M., Haluzík, M., Gay, S., & Šenolt, L. (2009). The role of resistin as a regulator of inflammation: Implications for various human pathologies. *Clinical immunology*, *133*(2), 157-170.
42. Follner, A., & Bodden, E. (2016). Ropocop—dynamic mitigation of code-reuse attacks. *Journal of Information Security and Applications*, *29*, 16-26.
43. Friedman, A., & Schuster, A. (2010). Data mining with differential privacy. In *Proceedings of the 16th ACM SIGKDD international conference on Knowledge discovery and data mining*, 3(1), 493-502.

44. Friedman, A., Wolff, R., & Schuster, A. (2008). Providing k-anonymity in data mining. *The VLDB Journal*, *17*(4), 789-804.

45. Fu, A. Y., Wenyin, L., & Deng, X. (2006). Detecting phishing web pages with visual similarity assessment based on earth mover's distance (EMD). *IEEE transactions on dependable and secure computing*, *3*(4), 301-311.

46. Giannotti, F., Lakshmanan, L. V., Monreale, A., Pedreschi, D., & Wang, H. (2012). Privacy-preserving mining of association rules from outsourced transaction databases. *IEEE Systems Journal*, *7*(3), 385-395.

47. Göktaş, E., Gawlik, R., Kollenda, B., Athanasopoulos, E., Portokalidis, G., Giuffrida, C., & Bos, H. (2016). Undermining information hiding (and what to do about it). In *25th {USENIX} Security Symposium ({USENIX} Security, 16,* 105-119.

48. Graziano, M., Balzarotti, D., & Zidouemba, A. (2016, May). ROPMEMU: A framework for the analysis of complex code-reuse attacks. In *Proceedings of the 11th ACM on Asia Conference on Computer and Communications Security*, 12(2), 47-58.

49. Gupta, S., & Gupta, B. B. (2016). XSS-SAFE: a server-side approach to detect and mitigate cross-site scripting (XSS) attacks in JavaScript code. *Arabian Journal for Science and Engineering*, *41*(3), 897-920.

50. Habibi, J., Gupta, A., Carlsony, S., Panicker, A., & Bertino, E. (2015). Mavr: Code reuse stealthy attacks and mitigation on unmanned aerial vehicles. In *2015 IEEE 35th International Conference on Distributed Computing Systems*, 3 (1), 642-652.

51. Herman, D. S., Lam, L., Taylor, M. R., Wang, L., Teekakirikul, P., Christodoulou, D., ... & Seidman, C. E. (2012). Truncations of titin causing dilated cardiomyopathy. *New England Journal of Medicine*, *366*(7), 619-628.

52. Hwang, K., Dave, P., & Tanachaiwiwat, S. (2003). NetShield: Protocol anomaly detection with datamining against DDoS attacks. In *Proceedings of the 6th International Symposium on Recent Advances in Intrusion Detection, Pittsburgh, 2(12),* 8-10.

53. Jabez, J., & Muthukumar, B. (2014). INTRUSION DETECTION SYSTEM: TIME PROBABILITY METHOD AND HYPERBOLIC HOPFIELD NEURAL NETWORK. *Journal of Theoretical & Applied Information Technology*, *67*(1), 4-10.

54. Jajodia, D. B. J. C. S., & Wu, L. P. N. (2001). Adam: Detecting intrusions by data mining. In *Workshop on Information Assurance and Security*, 1, 1100.
55. Kayaalp, M., Schmitt, T., Nomani, J., Ponomarev, D., & Ghazaleh, N. A. (2013). Signature-based protection from code reuse attacks. *IEEE Transactions on Computers*, *64*(2), 533-546.
56. Kerschbaum, F., Spafford, E. H., & Zamboni, D. (2002). Using internal sensors and embedded detectors for intrusion detection 1. *Journal of Computer Security*, *10*(1-2), 23-70.
57. Kumar, G. R., Singh, H. P., & Rajasekhar, N. (2019). Review Of Various Security Issues In Data Mining. *Turkish Journal of Physiotherapy and Rehabilitation*, *32*, 2.
58. Kumar, V., Chauhan, H., & Panwar, D. (2013). K-means clustering approach to analyze NSL-KDD intrusion detection dataset. *International Journal of Soft Computing and Engineering (IJSCE) ISSN*, 2231-2307.
59. Kumaraguru, P., Sheng, S., Acquisti, A., Cranor, L. F., & Hong, J. (2010). Teaching Johnny not to fall for phish. *ACM Transactions on Internet Technology (TOIT)*, *10*(2), 1-31.
60. Latah, M., & Toker, L. (2020). An efficient flow-based multi-level hybrid intrusion detection system for software-defined networks. *CCF Transactions on Networking*, *3*(3), 261-271.
61. Lee, Y. J. (2014). Privacy-preserving data mining for personalized marketing. *International Journal of Computer Communications and Networks (IJCCN)*, *4*(1), 3-19.
62. Li, Y., & Guo, L. (2007). An active learning based TCM-KNN algorithm for supervised network intrusion detection. *Computers & security*, *26*(7-8), 459-467.
63. Liu, X., Zhu, P., Zhang, Y., & Chen, K. (2015). A collaborative intrusion detection mechanism against false data injection attack in advanced metering infrastructure. *IEEE Transactions on Smart Grid*, *6*(5), 2435-2443.
64. Lu, R., Liang, X., Li, X., Lin, X., & Shen, X. (2012). EPPA: An efficient and privacy-preserving aggregation scheme for secure smart grid communications. *IEEE Transactions on Parallel and Distributed Systems*, *23*(9), 1621-1631.

65. Lu, W., & Traore, I. (2004). Detecting new forms of network intrusion using genetic programming. *Computational intelligence*, *20*(3), 475-494.

66. Marchal, S., François, J., State, R., & Engel, T. (2014). PhishStorm: Detecting phishing with streaming analytics. *IEEE Transactions on Network and Service Management*, *11*(4), 458-471.

67. Mitchell, R., & Chen, R. (2013). Adaptive intrusion detection of malicious unmanned air vehicles using behavior rule specifications. *IEEE transactions on systems, man, and cybernetics: systems*, *44*(5), 593-604.

68. Mitchell, R., & Chen, R. (2013). Effect of intrusion detection and response on reliability of cyber physical systems. *IEEE Transactions on Reliability*, *62*(1), 199-210.

69. Mitropoulos, D., Stroggylos, K., Spinellis, D., & Keromytis, A. D. (2016). How to train your browser: Preventing XSS attacks using contextual script fingerprints. *ACM Transactions on Privacy and Security (TOPS)*, *19*(1), 1-31.

70. Mohammed, N., Alhadidi, D., Fung, B. C., & Debbabi, M. (2013). Secure two-party differentially private data release for vertically partitioned data. *IEEE transactions on dependable and secure computing*, *11*(1), 59-71.

71. Mohanappriya, G., & Vijayalakshmi, D. (2018). Symmetric division degree index and inverse sum index of transformation graph. In *Journal of Physics: Conference Series*, 1139(1), 012048.

72. Musuvathi, M., Park, D. Y., Chou, A., Engler, D. R., & Dill, D. L. (2002). CMC: A pragmatic approach to model checking real code. *ACM SIGOPS Operating Systems Review*, *36*(SI), 75-88.

73. Nethravathi, P. C., Kumar, M. P., Suresh, D., Lingaraju, K., Rajanaika, H., Nagabhushana, H., & Sharma, S. C. (2015). Tinospora cordifolia mediated facile green synthesis of cupric oxide nanoparticles and their photocatalytic, antioxidant and antibacterial properties. *Materials Science in Semiconductor Processing*, *33*, 81-88.

74. Ni, M., Li, T., Li, Q., Zhang, H., & Ye, Y. (2016). FindMal: A file-to-file social network based malware detection framework. *Knowledge-Based Systems*, *112*, 142-151.

75. Parmar, G., & Mathur, D. K. (2015). Proposed Preventive measures and Strategies Against SQL injection Attacks. *Indian Journal of Applied Research*, *5*(5), 716-718.

76. Patel, D. K. B., & Bhatt, S. H. (2014). Implementnig Data Mining for Detection of Malware from Code. *Compusoft*, *3*(4), 732.

77. Pinkas, B. (2002). Cryptographic techniques for privacy-preserving data mining. *ACM Sigkdd Explorations Newsletter*, *4*(2), 12-19.

78. Rajasekaran, K., & Nirmala, K. (2015). A Novel and Advanced Data Mining Model Based Hybrid Intrusion Detection Framework. *TELKOMNIKA Indonesian Journal of Electrical Engineering*, *13*(2), 223-231.

79. Rasheed, F., & Alhajj, R. (2013). A framework for periodic outlier pattern detection in time-series sequences. *IEEE transactions on cybernetics*, *44*(5), 569-582.

80. Rauzy, P., & Guilley, S. (2014). A formal proof of countermeasures against fault injection attacks on CRT-RSA. *Journal of Cryptographic Engineering*, *4*(3), 173-185.

81. Roesch, M. (1999, November). Snort: Lightweight intrusion detection for networks. In *Lisa*, 99(1), 229-238.

82. Roglia, G. F., Martignoni, L., Paleari, R., & Bruschi, D. (2009, December). Surgically returning to randomized lib (c). In *2009 Annual Computer Security Applications Conference*, 22(1), 60-69.

83. Roopa, M. S., Pattar, S., Buyya, R., Venugopal, K. R., Iyengar, S. S., & Patnaik, L. M. (2019). Social Internet of Things (SIoT): Foundations, thrust areas, systematic review and future directions. *Computer Communications*, *139*, 32-57.

84. Roughan, M., & Zhang, Y. (2006). Secure distributed data-mining and its application to large-scale network measurements. *ACM SIGCOMM Computer Communication Review*, *36*(1), 7-14.

85. Samanthula, B. K., Elmehdwi, Y., & Jiang, W. (2014). K-nearest neighbor classification over semantically secure encrypted relational data. *IEEE transactions on Knowledge and data engineering*, *27*(5), 1261-1273.

86. Shapiro, S. S. (2012). Charleston, South Carolina 29403 USA. *Statistical Distributions in Scientific Work: Volume 5—Inferential Problems and Properties*, *79*, 25.

87. Siddiqui, M., Wang, M. C., & Lee, J. (2009). Detecting internet worms using data mining techniques. *Journal of Systemics, Cybernetics and Informatics*, *6*(6), 48-53.

88. Snow, K. Z., Monrose, F., Davi, L., Dmitrienko, A., Liebchen, C., & Sadeghi, A. R. (2013, May). Just-in-time code reuse: On the effectiveness of fine-grained address space layout randomization. In *2013 IEEE Symposium on Security and Privacy*, 3(1), 574-588.

89. Song, Q. H., & Hwang, K. C. (2007). Direct observation for photophysical and photochemical processes of folic acid in DMSO solution. *Journal of Photochemistry and Photobiology A: Chemistry*, *185*(1), 51-56.

90. Soni, P., & Sharma, P. (2014). An intrusion detection system based on KDD-99 data using data mining techniques and feature selection. *International Journal of Soft Computing and Engineering (IJSCE)*, *4*(3), 3-190.

91. Sunar, B., Martin, W. J., & Stinson, D. R. (2006). A provably secure true random number generator with built-in tolerance to active attacks. *IEEE Transactions on computers*, *56*(1), 109-119.

92. Tian, T., Wang, J., & Zhou, X. (2015). A review: microRNA detection methods. *Organic & biomolecular chemistry*, *13*(8), 2226-2238.

93. Tseng, C. Y., Balasubramanyam, P., Ko, C., Limprasittiporn, R., Rowe, J., & Levitt, K. (2003). A specification-based intrusion detection system for AODV. In *Proceedings of the 1st ACM workshop on Security of ad hoc and sensor networks,* 1(2), 125-134.

94. Vaidya, J., Shafiq, B., Fan, W., Mehmood, D., & Lorenzi, D. (2013). A random decision tree framework for privacy-preserving data mining. *IEEE transactions on dependable and secure computing*, *11*(5), 399-411.

95. Van Der Veen, V., Göktas, E., Contag, M., Pawoloski, A., Chen, X., Rawat, S., ... & Giuffrida, C. (2016, May). A tough call: Mitigating advanced code-reuse attacks at the binary level. In *2016 IEEE Symposium on Security and Privacy, 12(1),* 934-953.

96. Van Dongen, J. J. M., Langerak, A. W., Brüggemann, M., Evans, P. A. S., Hummel, M., Lavender, F. L., ... & Macintyre, E. A. (2003). Design and standardization of PCR primers and protocols for detection of clonal immunoglobulin and T-cell receptor gene recombinations in suspect lymphoproliferations: report of the BIOMED-2 Concerted Action BMH4-CT98-3936. *Leukemia*, *17*(12), 2257-2317.

97. Vasanthakumar, G. U., Prajakta, B., Shenoy, P. D., Venugopal, K. R., & Patnaik, L. M. (2015). PIB: Profiling influential blogger in online social networks, a knowledge driven data mining approach. *Procedia Computer Science*, *54*, 362-370.
98. Vasanthakumar, G. U., Sunithamma, K., Shenoy, P. D., & Venugopal, K. R. (2017). An overview on user profiling in online social networks. *Int. J. Appl. Inf. Syst*, *11*(8), 25-42.
99. Vittapu, M. S., Sunkari, V., & Abate, A. Y. (2015). The Practical Data Mining Model for Efficient IDS Through Relational Databases. *International Journal of Research in Engineering and Science*, *3*(1), 20-30.
100. Wu, B., Lu, T., Zheng, K., Zhang, D., & Lin, X. (2014). Smartphone malware detection model based on artificial immune system. *China Communications*, *11*(13), 86-92.
101. Wu, X., Kumar, V., Quinlan, J. R., Ghosh, J., Yang, Q., Motoda, H., ... & Steinberg, D. (2008). Top 10 algorithms in data mining. *Knowledge and information systems*, *14*(1), 1-37.
102. Xia, C., Hua, J., Tong, W., & Zhong, S. (2020). Distributed K-Means clustering guaranteeing local differential privacy. *Computers & Security*, *90*, 101699.
103. Yao, F., Chen, J., & Venkataramani, G. (2013). Jop-alarm: Detecting jump-oriented programming-based anomalies in applications. In *2013 IEEE 31st International Conference on Computer Design (ICCD)*, 4 (4), 467-470.
104. Yeh, C. H., Lee, G., & Lin, C. Y. (2015). Robust laser speckle authentication system through data mining techniques. *IEEE Transactions on Industrial Informatics*, *11*(2), 505-512.
105. Yeung, D. Y., & Ding, Y. (2003). Host-based intrusion detection using dynamic and static behavioral models. *Pattern recognition*, *36*(1), 229-243.
106. Zhu, T., Xiong, P., Li, G., & Zhou, W. (2014). Correlated differential privacy: Hiding information in non-IID data set. *IEEE Transactions on Information Forensics and Security*, *10*(2), 229-242

Chapter 3

Classification Approaches in Data Mining

CONTENTS

3.1. INTRODUCTION

A type of knowledge discovery that is used to solve issues in a particular domain is known as data mining. Taxonomy is a strategy for identifying unidentified data classes. Neural networks, bayesian, rule-based, decision trees, and other categorization methods exist for taxonomy. Unnecessary characteristics must be eliminated before the mining approach can be used. Various characteristic selection approaches, such as embedding, filter, and wrapper methodology, are used to filter data. This chapter serves as an introduction to the various categorization and characteristic selection strategies.

As even the world becomes more complicated, overwhelmed us with data, therefore data mining provides the only way to understand the patterns that lie behind it (Witten & Frank, 2002). As the amount of the data expands and the number of the dimension grows, the human method of data analysis gets laborious, necessitating the computerization of the process.

The automatic method of knowledge discovery through databases is referred to as Knowledge Discovery from Data (KDD). Data integration, Data cleansing, data selection, data mining, pattern assessment, data transformation, and representation of knowledge are all phases in the KDD process.

Data mining is a phase in the knowledge discovery process, that may be defined as the extraction or mining of knowledge from data huge volumes (Han et al., 2006). Data mining is a type of knowledge discovery that is used to solve issues in a particular domain. Data mining may alternatively be defined as a complex process that automatically extracts important concealed knowledge from data and presents it in the shape of patterns, concepts, and rules, etc. (Jun-Shan et al., 2003). The information gained by data mining enables the consumer to discover intriguing regularities and patterns hidden deep within the data to aid in judgment.

The tasks of data mining may be divided into two subgroups: predictive and descriptive. Descriptive mining activities describe the overall characteristics of the database's data. To create predictions, predictive mining activities draw inferences about existing data. The mining task may be split into four categories based on distinct goals: clustering analysis, association analysis, class/concept description, and classification or prediction (Beniwal & Arora, 2012).

The chapter gives an overview of the different classification strategies and feature selection that are employed in mining.

3.2. PREPROCESSING OF DATA

The data that can be mined is raw data. Data will be in many forms since it originates from various sources, and it may include irrelevant characteristics, missing data, noisy data, and so on. When employing any data mining method, the data must be pre-processed, that may be performed in the following ways (Alfano et al., 2011):

- Data Integration - When the data to be mined originates from many sources, it must be integrated, which entails eliminating discrepancies in characteristic value names or characteristic names among data sets from various sources.
- Data Cleaning — This phase may entail finding and fixing data mistakes, bringing in missing numbers, and so on. (Simoudis et al., 1994; Guyon et al., 2002) Discussing several data cleansing techniques below.
- Discretization - Whenever a data mining method can't handle continuous characteristics, it's time to use discretization. This phase entails converting a continuous characteristic to a categorical characteristic, that requires only a minimal discrete value. The comprehensibility of newly acquired knowledge is frequently improved by discretization (Wu, 1996; Gupta et al., 2010).
- Feature Selection - Because not all features are useful for mining, feature selection is necessary to choose a subset of essential features from all original features.

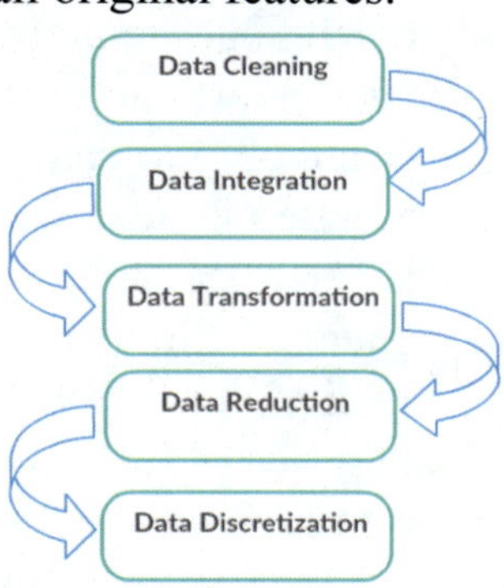

Figure 3.1. Common data processing stages

Source:https://www.researchgate.net/figure/Data-Preprocessing-Steps_fig1_291019609

3.3. SELECTION OF FEATURE

In the data that can be mined, there might be a lot of useless characteristics. As a result, they must be eliminated. In addition, many mining algorithms struggle with huge numbers of characteristics or qualities. As a result, feature selection approaches must be used even before the mining algorithm is used. The basic goals of selecting features are to minimize overfitting, create models that are quicker and less expensive, and enhance model performance (Ziarko, 1993).

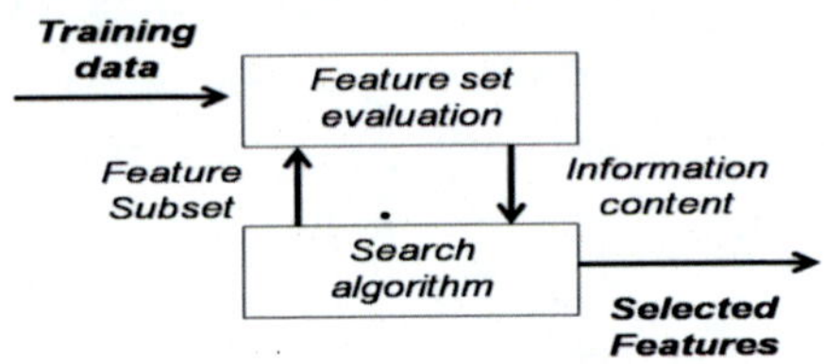

Figure 3.2. Data mining feature selection technique

Source: http://www.e2matrix.com/blog/2018/02/06/feature-selection-in-data-mining/

4A choosing of optimum characteristics adds an added layer of sophistication to the modeling because, rather than identifying optimal variables for the entire collection of characteristics, the model variables must first be optimized for the optimal characteristic subset (Shiba et al., 2003). Wrapper and Filter techniques are two types of feature selection strategies. The feature selection technique in the filter technique is autonomous of the data mining method to be used to the selected features and it evaluates the significance of characteristics by examining simply the data's intrinsic characteristics. A feature relevancy score is generated in most situations, and features with low scoring are eliminated. The classification algorithm is fed the subset of characteristics that remain after feature elimination. Filter methods have the benefits of being easy to scale to high-dimensional databases, being analytically simple and quick, and being free of the mining algorithm, this means characteristic selection only has to be done once, and then alternative classifiers may be assessed (Gupta et al., 2010). Filter methods have the drawback of ignoring the classifier's interaction, and also most suggested methods are univariate, meaning that every feature is kept separate, disregarding attribute dependencies, which may result in poor classification effectiveness when particularly in comparison to other kinds of attribute selection methods. A variety of multivariate filter approaches were proposed to solve the difficulty of disregarding feature interdependence, to incorporate characteristic dependencies to some extent. The model hypothesis discovery is embedded into the search of feature subset using wrapper techniques. The feature selection technique in the wrapper method uses the output of the data mining method to assess how excellent a particular feature subset is. A search process is specified in the space of potential subsets of features, and multiple subsets of attributes are created and assessed in this arrangement. The reliability of a feature subset is directly assessed by the effectiveness of the data mining method used to that attribute subset, which is a key feature of the wrapper method. Because the data mining method is executed for each feature subset examined by the search, the wrapper method is significantly quicker than the filter. Furthermore, the wrapper method becomes even more computationally costly if multiple distinct data mining techniques are to be implemented to the data (Raymer et al., 2000). Wrapper methods have the benefit of allowing model selection and feature subset search to interact, as well as the capability to account for feature interdependence. These approaches have the disadvantage of being more prone to overfitting than filter methods as well as being computationally expensive. Some other type of feature selection approach, known as embedded technique, was proposed, where the features optimal subset search is incorporated within the classifier building and may be viewed as a search in the joint space of attribute subsets and assumptions. Embedded methods, like wrappers, are so unique to a particular learning method. Embedded techniques offer the benefit of including the classification model's interaction while being considerably lower computationally demanding than wrapper techniques (Saeys et al., 2007).

3.4. CATEGORIZATION

There are three types of learning techniques for data mining algorithms: unsupervised, supervised, and semi-supervised.

The algorithm in supervised learning works with a collection of samples whose labels are specified. In the scenario of a categorization job, the labels could be nominal numbers, whereas, in the regression work case, the labels can be numerical numbers.

In non-supervised learning, on the other hand, the examples labels in the dataset are unidentified, and the method generally tries to group examples based on their attribute numbers' resemblance, which is referred to as a clustering job (Brameier & Wiuf, 2007).

Ultimately, partially-supervised learning is commonly employed when only a limited number of labeled examples are accessible, but a huge proportion of unlabeled examples are.

The categorization job may be thought of as a supervised approach in which every instance is assigned to a class based on the value of a variable.

or s unique objective. The aim property might have several category values, each of which corresponds to a different class. Every example is comprised of two parts: a collection of predictor feature values and a value for the target attribute. The former is being used to forecast the latter's value. The predictor characteristics should be useful in forecasting an instance's class. The collection of instances to be mined is split into two independently exhaustive and exclusive sets, known as the test set or the training set, in the categorization job (Hagras, 2004). The categorization process is separated into two different stages: training, which includes building a classification model using the testing set, and training, when involves evaluating the model on the test set. The method has accessibility to the numbers of both target and predictor characteristics across all examples within the training set throughout the training phase, and it utilizes this knowledge to create a classification model. This model contains classification information (basically, a connection among predictor attribute classes and values) that enables the class of an example to be predicted based on the predictor feature values. The test set does not display the examples class values for testing. Once a prediction is formed is the method permitted to observe the actual class of a just classified during the testing phase. One of the main aims of a categorization method is to improve the classification model's prediction accuracy when categorizing cases in the test set that were not encountered while training (Dembele & Kastner, 2003). A categorization method's knowledge can be described in a variety of ways, including decision trees, rules, and Bayesian networks. In the next part, we'll go through various categorization approaches.

3.5. CATEGORIZATION TECHNIQUES

The following are some examples of categorization techniques:

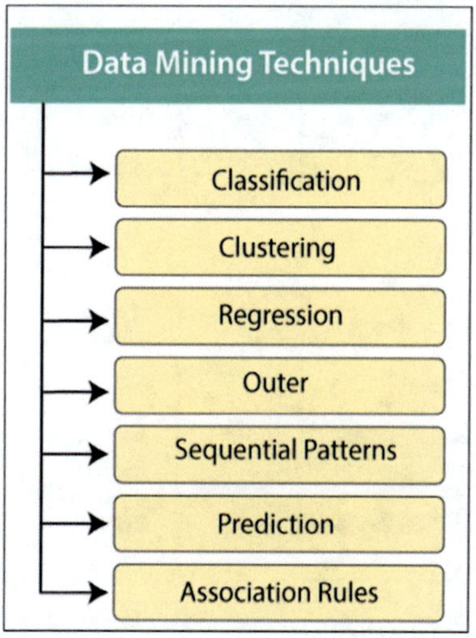

Figure 3.3. Data mining various techniques.

Source: https://www.javatpoint.com/data-mining-techniques

3.5.1. Classifiers Based on Rules

The exploration of easy-to-understand and high-level classifying rules of the context - if-then - is the focus of rule-based classifiers. The rules are divided into two sections, the rule consequent, and the rule antecedent. The rule-antecedent, also known as the if-part, defines a set of parameters referring to predictor attribute-value, while the rule-consequent, also known as the then-part, defines the category anticipated by the rule for any instance that meets the rule antecedent's conditions. Different classification algorithms can be used to create these rules, the most well-known of which are sequential-covering rule induction-algorithm and decision-tree induction-algorithm (Cerri et al., 2015).

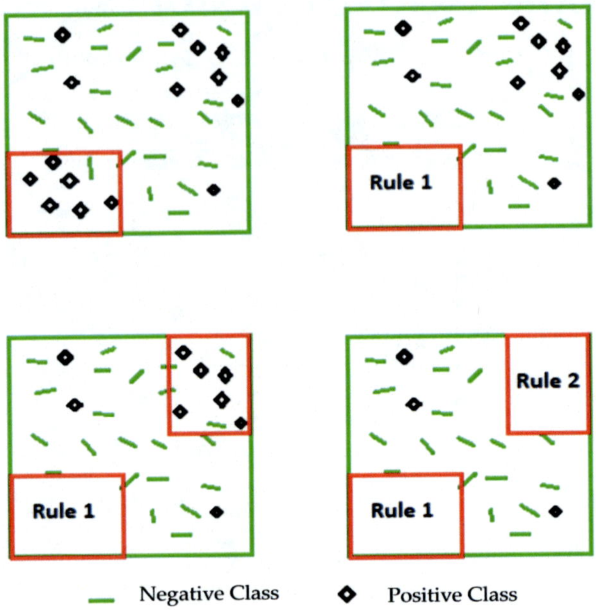

Figure 3.4. Generation of Sequential Rule

Source: https://www.geeksforgeeks.org/rule-based-classifier-machine-learning/

3.5.2. Bayesian Networks

Bayesian network - BN - is made up of a probability distribution, acyclic graph, and a directed for each node based on nearest predecessors (Darwiche et al., 2009). A Bayes-Network Classifier is built based on a Bayesian network, that represents a joint-probability distribution over a series of categorical features. The columns are made up of two parts: the conditional-probability and a directed acyclic graph (G) with arcs and nodes. The nodes illustrate attributes, while the arcs reflect direct relationships. One indicator of a BN's complexity is the density of its arcs. Simplified probabilistic models - such as hidden-Markov models and naive Bayes - can be represented by sparse BNs, whereas dense BNs can obtain extremely complex-model. As a result, BNs offer a versatile method for probabilistic modeling (Cooper et al., 2010).

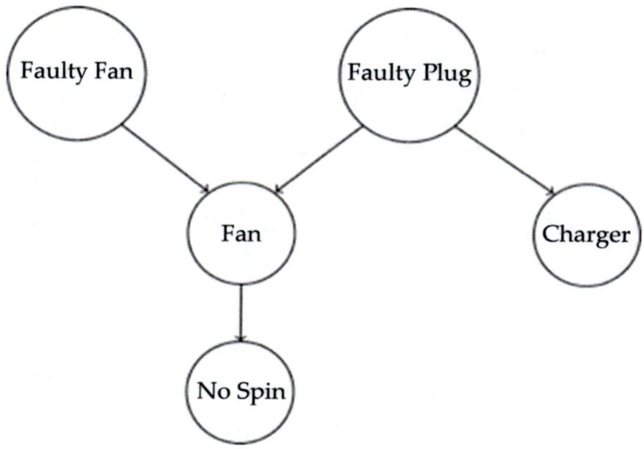

Figure 3.5. An example of a basic Bayesian network

Source: https://www.sciencedirect.com/topics/mathematics/bayesian-network

3.5.3. Decision Tree

The Decision-Tree Category is made up of a decision tree that is obtained from situations. Leaf nodes along with the internal and root nodes are made up of the decision tree. Attributes are related to the internal and root nodes, while classes are affiliated with the leaf nodes. In essence, every non-leaf node has a departing offshoot for each potential value of the node's attribute. Starting with the root, subsequent internal nodes are attended until a leaf node is approached to predict the class for a new observation that used a decision tree. A test is run at each internal node and the root node. The next node and the section traversed attended are determined by the test's results. The instance's class is the category of the last leaf node (Garofalakis et al., 2003).

3.5.4. Nearest Neighbour

All incidences in a Nearest-Neighbor Classification are assumed to correlate to positions in n-dimensional orbit. All observations are acknowledged during the learning process. Whenever a new point is classified, the nearest points are observed and used, along with a weight, to determine the new point's class value. Closer points are given higher weights to improve the accuracy (Mitchell, 1997).

3.5.5. Artificial Neural-Network

An artificial neural network commonly referred to as a neural network, is a computational and mathematical model based on biological neural networks, or, to put it another way, modeling of a biological neural system. During the process of learning, an ANN is typically a resilient system that alters its structure based on internal and external information that flows via the network (Singh & Chauhan, 2009). A Neural-Network Classification model is made up of interconnected neurons in a neural network. A neuron receives negative and positive stimuli - numerical values - from other nerve cells and stimulates itself when the summation of the stimuli is larger than a predefined threshold value. The summation of stimuli is usually transformed non-linearly by the neuron's output value. The non-linear transformation is adjusted by some linear functions in more sophisticated models.

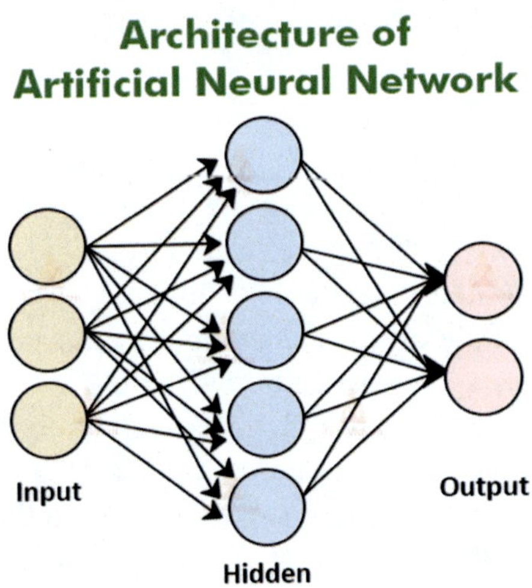

Figure 3.6. The architecture of Artificial Neural Networks

Source: https://techvidvan.com/tutorials/artificial-neural-network/

3.5.6. Support vector machines

Support Vector Machines (Vapnik, 1999) (SVMs) are binary classification techniques in their most basic form. SVM is a statistical learning theory-based classification system. It's been used successfully in areas like bio-sequence analysis, text categorization, image classification, and hand-written character recognition, among others. While the SVM segregates the categories with a decision surface that maximizes the distance between the categories. The data points closest to the hyper-plane are referred to as support vectors and the optimal-hyperplane is often referred to as the surface. The practicing set's essential aspects are the support vectors. The kernel function is the method that determines how the mapping is done. Using nonlinear kernels, the SVM can be transformed into a non-linear classifier. By combining numerous binary SVM-classifier, it can be used to classify multiple classes (Suykens & Vandewalle, 1999). The judgemental values of every pixel for each class are the outcome of SVM classification, and they are used to evaluate likelihood. The probabilistic values illustrate - true - probability in the context that they all fall between 0 and 1, and the summation of these values for every pixel equals 1. The highest probability is then chosen for categorization. The penalty variable in SVM enables some misclassification, which is extremely crucial for non-separable training models. The penalty variable regulates the trade-off between imposing strict margins and allowing training error. It produces a smooth margin that allows for some misclassifications, like training points on the incorrect facet of the hyper-plane. Boosting the penalty parameter's value increases the price of misclassifying places and pushes the formation of a more precise model, which may be less generalizable (Chang et al., 2000).

3.5.7. Rough Sets

A primary set is any collection of indistinguishable - similar - objects. A precise and crisp set is any confederation of some basic sets; alternatively, the set is rough - vague and imprecise. Every rough-set has demarcation instances or objects that cannot be categorized as candidates of the set or its supplement with surety using the existing information (Brunato & Battiti, 2005). Certainly, unlike precise sets, rough sets cannot be described in terms of knowledge about their components. A pair of precise-set called the upper and lower estimation of the rough-set are correlated with any rough-set. The lowest approximation includes all items that must correspond to the set, whereas the higher approximation includes all objects that may correspond to the set. The rough set's border area is defined by the disparity among the upper approximations and lower. The rough set strategy to data analysis seems to have several significant benefits, including efficient methodologies for discovering hidden patterns within the data, identifying relationships that would not be discovered utilizing statistical techniques, allowing both quantitative and qualitative data, discovering minimal data sets, evaluating data significance, and being simple to comprehend (Pawlak, 1982).

3.5.8. Fuzzy Logic

The logic of fuzzy is a multivalued logic that differs from "crisp logic," which is based on binary sets with two values. A truth value for fuzzy logic parameters somewhere between 1 and 0. it is a superset of Boolean logic which has been expanded to encompass partial truth. The membership function (MF) is indeed a curve that specifies how every point is translated to a membership value (or membership degree) on the input space ranging from 0 to 1. Type 1 and 2 fuzzy logic are the two types of fuzzy logic (Sekhar et al., 2013). The constant numbers are contained in Type 1. Type-2 is a type of Type 1 is where the fuzzy sets are derived from the current Type 1 Fuzzy Logic. The membership grades in a type-2 fuzzy set are also fuzzy. Every subset of the primary membership can indeed be classified as Type-2. A secondary membership exists for each main membership, which determines the primary membership's possibilities. Type-1 Fuzzy Logic can't deal with rule inconsistency. Type-2 Fuzzy Logic is capable of efficiently and effectively dealing with rule uncertainty (Tari et al., 2009). IF-THEN rules characterize Type-2 Fuzzy sets once again (Karnik et al., 1999). Since type reduction is time-consuming, Type-2 Fuzzy takes a lot of time to compute. Type-2 fuzzy is a type of fuzzy that may best define imprecision and uncertainty. "Fuzzy fuzzy" refers to type-2 fuzzy sets since the membership degree of fuzzy is also fuzzy, as a consequence of Type 1 fuzzy sets.

3.5.9. Genetic algorithms

GAs - Genetic Algorithms - are searching algorithms derived from natural genetics which provide rigorous search functions in complicated spaces, making them a viable solution to problems that require effective and efficient search methods (Woolley et al., 2011). It is an iterative method that works with a population - a group of potential solutions. Every solution is derived through de-coding and encoding strategies that allow us to visualize the solution as a genetic code and vise - versa. Originally, the population is created at random. A fitness value is assigned to each member of the population via a fitness function that represents their quality in fixing the specific issue. A fitness function evaluates a chromo-some to verify the solution's quality, or how efficient it is at fixing the issues. The chromo-some is the input to the fitness function, and the fitness-value of this chromosome is the outcome. Each loop determines the fitness of every candidate solution. The next step is a selection, which involves creating a temporary populace wherein the fittest candidates have a better likelihood of being used as parents for the succeeding generation than those who are less fit. candidates ls in this population are subjected to genital operators such as mutation and crossover, resulting in a new populace (Booker et al., 1989).

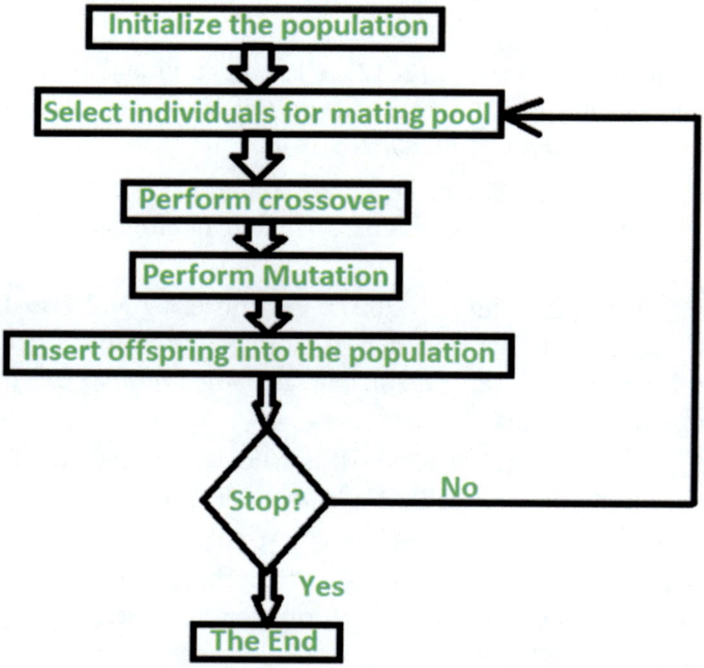

Figure 3.7. A basic genetic algorithm

Source: https://www.geeksforgeeks.org/encoding-methods-in-genetic-algorithm/

REFERENCES

1. Alfano, F., Bonadonna, C., Volentik, A. C., Connor, C. B., Watt, S. F., Pyle, D. M., & Connor, L. J. (2011). Tephra stratigraphy and eruptive volume of the May, 2008, Chaitén eruption, Chile. *Bulletin of Volcanology*, *73*(5), 613-630.
2. Beniwal, S., & Arora, J. (2012). Classification and feature selection techniques in data mining. *International journal of engineering research & technology (ijert)*, *1*(6), 1-6.
3. Booker, L. B., Goldberg, D. E., & Holland, J. H. (1989). Classifier systems and genetic algorithms. *Artificial intelligence*, *40*(1-3), 235-282.
4. Brameier, M., & Wiuf, C. (2007). Co-clustering and visualization of gene expression data and gene ontology terms for Saccharomyces cerevisiae using self-organizing maps. *Journal of biomedical informatics*, *40*(2), 160-173.
5. Brunato, M., & Battiti, R. (2005). Statistical learning theory for location fingerprinting in wireless LANs. *Computer Networks*, *47*(6), 825-845.
6. Cerri, R., Pappa, G. L., Carvalho, A. C. P., & Freitas, A. A. (2015). An extensive evaluation of decision tree–based hierarchical multilabel classification methods and performance measures. *Computational Intelligence*, *31*(1), 1-46.
7. Chang, C. C., Hsu, C. W., & Lin, C. J. (2000). The analysis of decomposition methods for support vector machines. *IEEE Transactions on Neural Networks*, *11*(4), 1003-1008.
8. Cooper, G. F., Hennings-Yeomans, P., Visweswaran, S., & Barmada, M. (2010). An efficient Bayesian method for predicting clinical outcomes from genome-wide data. In *AMIA Annual Symposium Proceedings,* 2010, 127.
9. Darwiche, F. Ž., Ugradar, S. T., & Turner, T. (2009). Junior doctors' knowledge and practice of electrocardiographic monitoring for high-risk patients receiving antipsychotic medications. *Psychiatric Bulletin*, *33*(10), 377-380.
10. Dembele, D., & Kastner, P. (2003). Fuzzy C-means method for clustering microarray data. *bioinformatics*, *19*(8), 973-980.
11. Garofalakis, M., Hyun, D., Rastogi, R., & Shim, K. (2003). Building decision trees with constraints. *Data Mining and Knowledge Discovery*, *7*(2), 187-214.

12. Gupta, A., Mehrotra, K. G., & Mohan, C. (2010). A clustering-based discretization for supervised learning. *Statistics & probability letters*, *80*(9-10), 816-824.

13. Guyon, I., Weston, J., Barnhill, S., & Vapnik, V. (2002). Gene selection for cancer classification using support vector machines. *Machine learning*, *46*(1), 389-422.

14. Hagras, H. A. (2004). A hierarchical type-2 fuzzy logic control architecture for autonomous mobile robots. *IEEE Transactions on Fuzzy systems*, *12*(4), 524-539.

15. Han, J., Kamber, M., & Mining, D. (2006). Concepts and techniques. *Morgan Kaufmann*, *340*, 94104-3205.

16. Huang, D., & Pan, W. (2006). Incorporating biological knowledge into distance-based clustering analysis of microarray gene expression data. *Bioinformatics*, *22*(10), 1259-1268.

17. Hurtado, J. E. (2004). An examination of methods for approximating implicit limit state functions from the viewpoint of statistical learning theory. *Structural Safety*, *26*(3), 271-293.

18. Jun-Shan, T., Wei, H., & Yan, Q. (2009). Application of genetic algorithm in data mining. In *2009 First International Workshop on Education Technology and Computer Science*, 2, 353-356.

19. Karnik, N. N., Mendel, J. M., & Liang, Q. (1999). Type-2 fuzzy logic systems. *IEEE transactions on Fuzzy Systems*, *7*(6), 643-658.

20. Martínez, L. G., Licea, G., Rodríguez, A., Castro, J. R., & Castillo, O. (2013). Using MatLab's fuzzy logic toolbox to create an application for RAMSET in software engineering courses. *Computer Applications in Engineering Education*, *21*(4), 596-605.

21. Mitchell, T. M. (1997). Artificial neural networks. *Machine learning*, *45*, 81-127.

22. Nijssen, S., & Fromont, E. (2010). Optimal constraint-based decision tree induction from item set lattices. *Data Mining and Knowledge Discovery*, *21*(1), 9-51.

23. Paul, A. K., & Shill, P. C. (2018). Incorporating gene ontology into fuzzy relational clustering of microarray gene expression data. *Biosystems*, *163*, 1-10.

24. Pawlak, Z. (1982). Rough sets. *International journal of computer & information sciences*, *11*(5), 341-356.

25. Rahmani, Z., Blouin, J. L., Creau-Goldberg, N., Watkins, P. C., Mattei, J. F., Poissonnier, M., ... & Aurias, A. (1989). Critical role of the D21S55 region on chromosome 21 in the pathogenesis of Down syndrome. *Proceedings of the National Academy of Sciences*, *86*(15), 5958-5962.
26. Rastogi, R., & Shim, K. (2000). PUBLIC: A decision tree classifier that integrates building and pruning. *Data Mining and Knowledge Discovery*, *4*(4), 315-344.
27. Raymer, M. L., Punch, W. F., Goodman, E. D., Kuhn, L. A., & Jain, A. K. (2000). Dimensionality reduction using genetic algorithms. *IEEE transactions on evolutionary computation*, *4*(2), 164-171.
28. Saeys, Y., Inza, I., & Larranaga, P. (2007). A review of feature selection techniques in bioinformatics. *bioinformatics*, *23*(19), 2507-2517.
29. Sekhar, K. R., Kalyan, V. S., & Kumar, B. P. (2013). Training of artificial neural networks in data mining. *International Journal of Innovative Technology and Exploring Engineering (IJITEE), ISSN*, 2278-3075.
30. Sharma, J., Chawla, S., & Dalhotra, S. (2013). A research agenda on artificial neural network topologies & data mining in neural network. *International Journal of Data & Network Security*, *1*, 41-47.
31. Shiba, O. A., Sulaiman, M. N., Ahmad, F., & Mamat, A. (2003). An experimental evaluation of case slicing as a new classification technique. *Journal of Information and Communication Technology*, *2*(2), 105-117.
32. Simoudis, E., Livezey, B., & Kerber, R. (1994). Integrating inductive and deductive reasoning for database mining. In *Proceedings of the 3rd International Conference on Knowledge Discovery and Data Mining*, 1 (2), 37-48.
33. Singh, Y., & Chauhan, A. S. (2009). NEURAL NETWORKS IN DATA MINING. *Journal of Theoretical & Applied Information Technology*, *5*(1).
34. Suykens, J. A. K., Lukas, L., Van Dooren, P., De Moor, B., & Vandewalle, J. (1999). Least squares support vector machine classifiers: a large scale algorithm. In *European Conference on Circuit Theory and Design*, 99, 839-842.
35. Suykens, J. A., & Vandewalle, J. (1999). Least squares support vector machine classifiers. *Neural processing letters*, *9*(3), 293-300.

36. Tang, Y., Zhang, Y. Q., Huang, Z., Hu, X., & Zhao, Y. (2008). Recursive fuzzy granulation for gene subsets extraction and cancer classification. *IEEE Transactions on Information Technology in Biomedicine*, *12*(6), 723-730.
37. Tari, L., Baral, C., & Kim, S. (2009). Fuzzy c-means clustering with prior biological knowledge. *Journal of biomedical informatics*, *42*(1), 74-81.
38. Vapnik, V. N. (1999). An overview of statistical learning theory. *IEEE transactions on neural networks*, *10*(5), 988-999.
39. Witten, I. H., & Frank, E. (2002). Data mining: practical machine learning tools and techniques with Java implementations. *Acm Sigmod Record*, *31*(1), 76-77.
40. Woolley, R. A., Stirling, J., Radocea, A., Krasnogor, N., & Moriarty, P. (2011). Automated probe microscopy via evolutionary optimization at the atomic scale. *Applied Physics Letters*, *98*(25), 253104.
41. Wu, X. (1996). A Bayesian discretizer for real-valued attributes. *The Computer Journal*, *39*(8), 688-691.
42. Yarveicy, H., Moghaddam, A. K., & Ghiasi, M. M. (2014). Practical use of statistical learning theory for modeling freezing point depression of electrolyte solutions: LSSVM model. *Journal of Natural Gas Science and Engineering*, *20*, 414-421.
43. Ziarko, W. (1993). Variable precision rough set model. *Journal of computer and system sciences*, *46*(1), 39-59.

Chapter 4

Application of Secure Data Mining in Fraud Detection

CONTENTS

4.1. INTRODUCTION

The paper describes the application of secure data mining methods to fraud examination. We demonstrate certain classification and forecast secure data mining methods which we consider significant for handling the detection of fraud. There occur several secure data mining algorithms and we show rule-based algorithms, result tree-based algorithms, and statistics-based algorithms. We demonstrate the Bayesian classification model for fraud detection in the insurance of automobiles. Naïve Bayesian visualization is chosen to examine and understand the expectations of the classifier. We demonstrate how ROC curves could be positioned for the assessment of the model to give a more instinctive examination of the models.

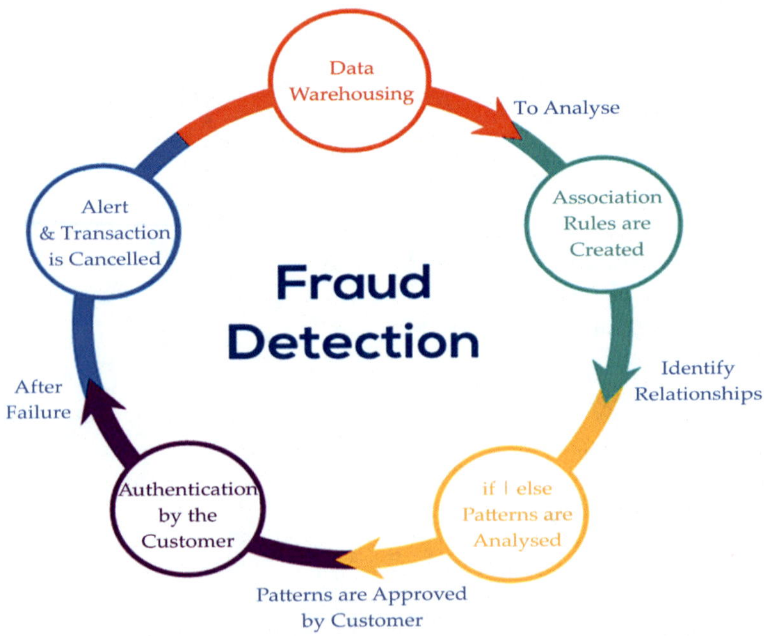

Figure 4.1. Detection of Fraud algorithm mechanisms

Source: https://www.omnisci.com/technical-glossary/fraud-detection-and-prevention

Secure data mining mentions to mining or extracting information from a huge amount of data. There are several secure data mining methods like regression, neural networks, clustering, and several predictive models. Now, we debate only a few methods of secure data mining which would be deliberated significant for handling detection of fraud. They are i) Bayesian network, for categorizing risk group, and ii) Decision tree, for forming an expressive model of every risk group.

Secure data Mining is related to (a) supervised learning founded on training data of recognized fraud and valid cases and (b) *unsupervised learning* having data that is not considered to be valid and fraud. Bedford's law could be taken as an instance of unsupervised learning (Bolton et al. 2002). The direct use of these techniques in forensic accounting is restricted because of the nearly complete absence of huge sets of fraud training information (Jensen, 1997; Bolton et al. 2002).

Telecommunications fraud, credit card fraud, checked forgery and insurance fraud are some of the major kinds of fraud. Insurance fraud is usual in travel, automobile. The NAIC Antifraud Task Force 2003 implemented "The Uniform Suspected Insurance Fraud Reporting Form", which substituted the previous Task Force form. This form regulates the information of insurance fraud for the insurance industry and forms it convenient to report and locate. Fraud detection contains three kinds of offenders (Baldock, 1997): i) Criminal offenders, ii) planned criminal offenders who are liable for main fraud, and iii) offenders who make fraud (termed soft fraud) when suffering from monetary poverty. Soft fraud is the toughest to decrease since the cost for every suspected instance is generally higher than the price of the fraud (National White Collar Crime Center, 2003). Kinds i) and ii) offenders, termed hard fraud, evade anti-fraud methods (Sparrow, 2002).

We present secure data mining methods which are most suitable for fraud examination. We give an instance of automobile insurance. Three secure data mining methods utilized for fraud examination are i) Decision tree, ii) backpropagation and iii) Bayesian network. Decision trees are utilized to make descriptive models. Descriptive models are formed to explain the features of fault. Bayesian network is the method utilized for task classification. Classification provides a set of predefined definite classes, defines which of these classes precise data fits to.

4.2. EXISTING FRAUD DETECTION SYSTEMS

An uncertain logic scheme (Altrock et al. 1995) integrated the real fraud assessment policy utilizing optimal threshold standards. The outcome presented the probabilities of fraud and the causes of why an insurance claim is fake. This scheme projected slightly better outcomes than the checkers. Additional logic system (Cox et al. 1995) utilized two methods to copy the fraud professionals reasoning, i) the discovery model utilizes an unverified neural network to discover the associations in data and to locate clusters, then patterns inside the clusters are recognized, and ii) the uncertain irregularity detection model, which utilized Wang-Mendel algorithm to locate how health care workers did fraud against the insurance companies. The EFD scheme (Major et al. 1995) incorporated the expert information with statistical data to recognize workers whose behavior didn't fit the rule.

The hot spots technique (Williams et al. 1997) done a three-step procedure: i) k-means clustering algorithm for the detection of the cluster is utilized due to the further clustering algorithms inclined to be computationally costly wherever the datasets are very huge, ii) C4.5 algorithm, the resultant decision tree could be transformed to a rule set and clipped, and iii) visualization tools for the assessment of rule, creating statistical summaries of the entities related with every rule. (Williams, 1999) prolonged the hot spots method to utilize genetic algorithms to produce and discover the rules.

The model of credit card fraud (Groth et al. 1998) proposed a classification method with legal/fraud characteristics, also a clustering followed through a classification method with no legal/fraud characteristic. Kohonen's Self-Organizing Feature Chart (Brockett et al. 1998) was utilized to classify automobile injury claims relying on the size of fraud doubt. The validity of the Feature Chart was then assessed by utilizing a backpropagation algorithm and nourish forward neural networks. The outcome presented that the technique was more dependable and consistent likened to the fraud assessment.

The methods of classification had proved to be very efficient in the detection of fraud (He et al. 1998; Chen et al. 1999) and thus, could be applied for the categorization of crime records. The dispersed data mining model (Chen et al. 1999) utilizes an accurate cost model to assess CART, naïve Bayesian classification models, and C4.5. The technique was implemented for transactions of a credit card. The neural data mining method (Brause et al. 1999) utilizes rule-based relation rules to extract symbolic data and to extract analog data it utilizes Radial Base Function neural network. The method debates the significance of the usage of non-numeric data in the detection of fraud. It was noticed that the outcomes of association rules enhanced the predictive exactness.

The SAS Enterprise Miner Software (SAS e-intelligence, 2000) relies on cluster detection, association rules, and classification methods for the detection of fraudulent claims. The ANN and (Artificial Neural Network) and BBN (Bayesian Belief Network) study utilized the STAGE algorithm for Bayesian Belief Network in the detection of fraud and backpropagation for Artificial Neural Network (Maes et al. 2002). STAGE continually substitutes amongst two stages of search: operating the original search technique on the objective function, and operating hill-climbing to enhance the value function. The outcome displays that Bayesian Belief Network was much quicker to train, however, were slower when used to new instances. FraudFocus Software (Magnify, 2002) mechanically scores entire claims. The scores are arranged in descending patterns of fraud potential and produce descriptive rules for claims of fraud. FairIsaac (Weatherford et al. 2002) suggested backpropagation neural networks for the utilization of fraudulent credit cards. The ASPECT group (Weatherford et al. 2002) is attentive to neural networks for training existing user profiles and histories of user profiles. A caller's present profile and the history of the profile are compared to find possible fraud. (Cahill et al. 2002) formed on the adaptive fraud detection outline (Fawcett et al. 1997) through implementing an event-driven method of conveying fraud scores for detection of fraud. The (Cahill et al. 2002) framework could also identify kinds of fraud by rules utilization. This framework had been utilized in both wired and wireless fraud detection schemes. (Ormerod el al. 2003) utilized dynamic Bayesian Belief Network termed Mass Detection tool for detection of fraudulent claims, which then utilized a rule generator termed Suspicion Building Tool.

The diverse kinds of fraud detection are insurance, internal, telecommunications, and credit card fraud detection. Detection of internal fraud comprises determining to report of fraudulent financial by management (Lin et al. 2003; Bell et al. 2000), and irregular retail transactions reporting through employees (Kim et al. 2003). There are 4 kinds for detection of insurance fraud: crop insurance (Little et al. 2002), health insurance, home insurance (Bentley, 2000; Von Altrock, 1997), and detection of automobile insurance fraud. A particular meta classifier(Phua et al. 2004) is utilized to chose the finest base classifiers, and then joined with these base classifiers' guess to enhance cost savings (stacking bagging). Detection of Automobile insurance fraud records set was utilized to demonstrate the issue of stacking-bagging (Burge et al. 2001; McGibney et al. 2003). Detection of Credit card fraud mentions screening credit applications, or/and logged transactions of credit cards. Telecommunications subscription records, and/or wireless and wired phone calls are checked. Detection of credit transactional fraud had been given by (Foster et al. 2004) and forecast of bad debts (Ezawa et al. 1996). Retail/ employee, national insurance of crop, and credit application (Wheeler et al. 2000; Little et al. 2002; Kim et al. 2003). The Literature emphasis on IP-based telecommunication services and video-on-demand websites Online buyers, and online sellers (Bhargava et al. 2003) could be checked through automated systems. Detection of fraud in government organizations like customs and tax had also been stated (Barse et al. 2003; Sherman, 2002).

We debated under supervised data mining methods for detecting crime utilizing Bayesian Belief Networks, Artificial Neural Networks, and Decision trees (Bonchi et al. 1999; Shao et al. 2002).

4.2.1. Bayesian Belief Networks

Bayesian Belief Networks gives a graphic model of fundamental relationships in which class membership prospects (Han et al. 2000) are projected so that a specific example is a fraud or legal (Prodromidis, 1999).

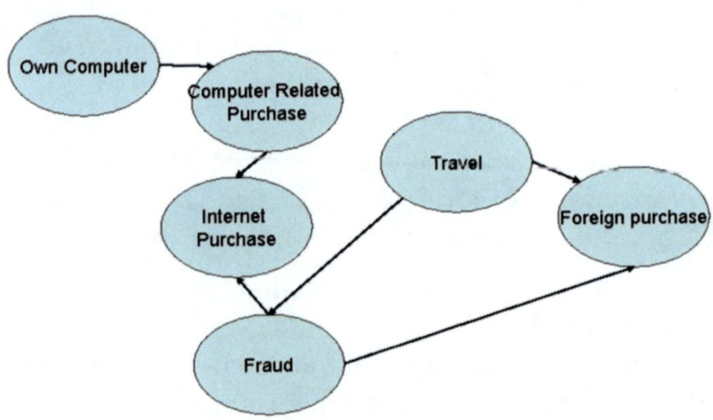

Figure 4.2. Extended Bayes network for detection of fraud

Source: https://slidetodoc.com/practice-of-bayesian-networks-data-mining-lab-4/

Naïve Bayesian classification supposes that the characteristics of an example are autonomous, provided the target attribute (Feelders et al. 2003). The purpose is to allow a novel instance to the class that had the maximum posterior possibility. The algorithm is very efficient and could provide better predictive exactness when compared to backpropagation and C4.5 decision trees (Domingos et al. 1996; Elkan et al. 2001). Though, when the features are redundant, the predictive precision is decreased (Witten et al. 1999).

4.2.2. Decision Trees

Decision trees are machine learning methods that direct a dependent attribute and independent attributes in a structure tree-shaped that signifies a set of decisions (Witten et al. 1999). Rules of classification, mined from decision trees, are IF-THEN terms in which the requirements are logically added and entire tests had to succeed if every rule is to be produced.

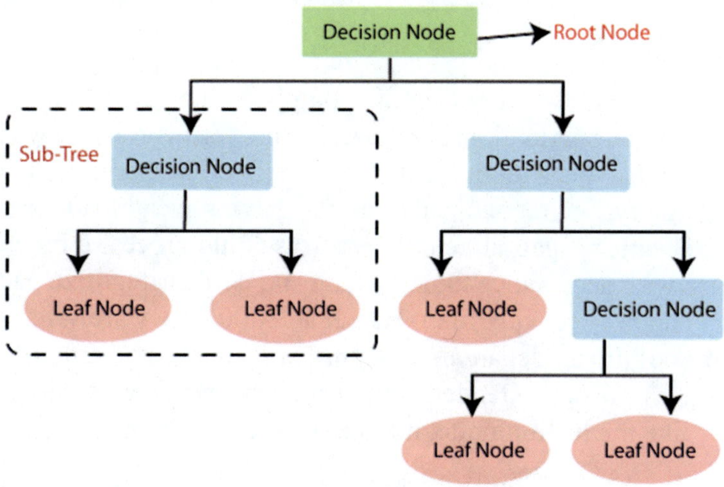

Figure 4.3. A simple decision tree for detection of fraud

Source: https://www.javatpoint.com/machine-learning-decision-tree-classification-algorithm

The associated applications comprise the analysis of examples from customs declaration fraud (Shao et al. 2002), governmental financial transactions (Mena et al. 2003), and drug smuggling to more severe crimes like as serial sex crimes (SPSS, 2003), homeland security (James et al. 2002; Mena et al. 2003) and drug-associated homicides. Secure data mining techniques had solved security and detection of criminal issues [Mena, 2003] gone through the subject, (link analysis, decision trees, intelligent agents, machine learning, text mining, neural networks, and self-organizing maps) for counter-intelligence agents, law enforcement investigators, security managers, information security analysts, and fraud specialists. The C4.5 (Quinlan et al. 1993) is utilized to split data into segments founded and to produce descriptive classification rules that could be utilized to classify a new example. The C4.5 could assist to make forecasts and mining crime patterns. It produces rules from trees (Witten et al.., 1999) and handles numeric features, pruning, approximating error rates, and missing values. The C4.5 executes marginally better than ID3 and CART (Prodromidis, 1999) in terms of predictive exactness. The learning and categorization steps are usually fast (Han et al. 2000). Though, the decline in performance could happen when C4.5 is implemented to huge datasets. The C5.0 shows marginal enhancements for induction of the decision tree.

4.2.3. Artificial Neural Networks

The Artificial Neural Networks signify complicated mathematical equations with exponentials, summations, and parameters for neurons copy (Berry et al. 2000). They had been implemented to classify crime examples like sexual offenses, burglary, and recognized criminals having facial features (Mena et al. 2003b). Backpropagation neural networks could process a huge number of examples with acceptance to noisy data and could categorize patterns on which they had not been qualified (Han et al. 2000). They are suitable where the outcomes of the model are more significant (Berry et al. 2000). Though, backpropagation needs extended training hours, extensive testing, holding parameters like the learning rate, number of secret neurons (Bigus, 1996).

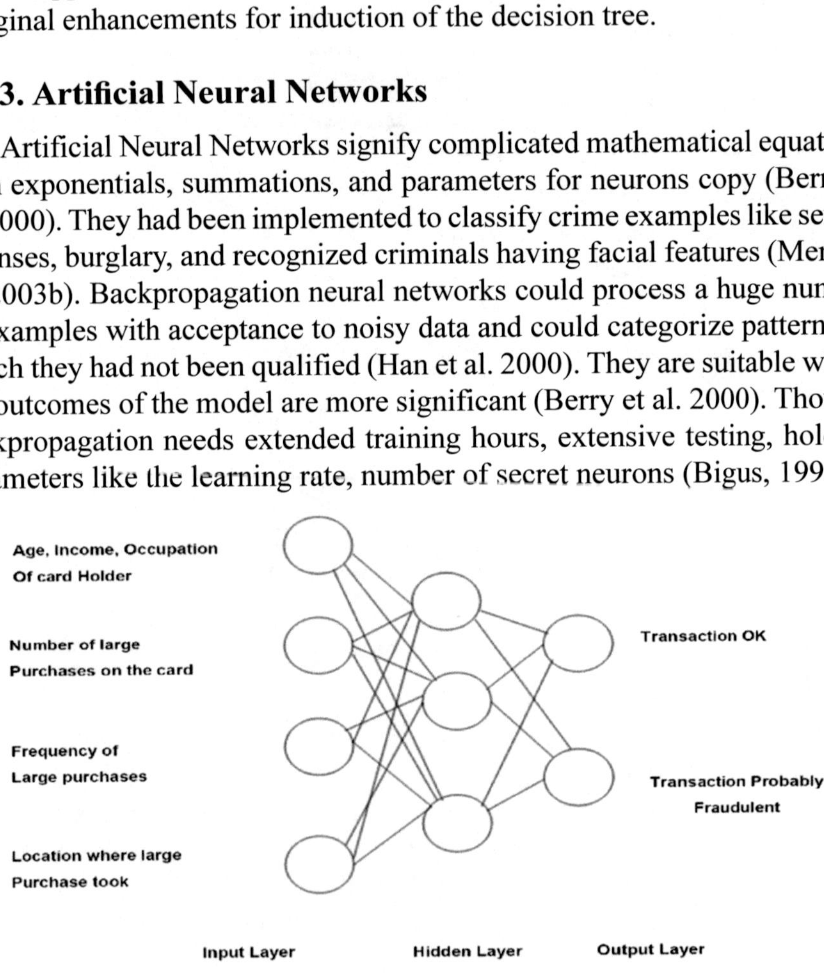

Figure 4.4. Layers of neural network in fraud detection of ATM card

Source: https://www.researchgate.net/figure/Layers-of-neural-network-in-ATM-card-fraud-detection_fig1_334898568

4.3. APPLICATIONS

The crime detection steps are i) classifiers, ii) incorporate multiple classifiers, iii) ANN method to clustering, and iv) visualization methods to explains the patterns.

4.3.1. Bayesian Network

The Bayesian Network is a focused acyclic graph, where every node signifies a random variable and is related with the conditional possibility of the node specified its parents. This model presents every variable in a specified domain as the dependences and node in the graph amongst these variables as arcs linking to the relevant nodes. This is, entire edges in the graphical model are focused and there are not any cycles (Riedinger et al. 2002; Yamanishi et al. 2004).

For the fraud detection purpose, we create two Bayesian networks to explain the auto insurance behavior. Initially, the Bayesian network is created to model behavior under the supposition that the driver is fake (F) and an additional model under the supposition the driver is a genuine user (NF), see Figure 3. The 'fraud net' is used by utilizing expert information. The 'user net' is used through utilizing records from drivers of non-fraudulent. Throughout the operation, the user net is revised to a precise user based on developing records. Through implanting proof in these networks (the perceived user behavior x derived from his toll tickets) and spreading it by the network, we could achieve the possibility of the measurement x in the two above-stated hypotheses. That means we get decisions as to what step an observed user behavior comes across as usual non-fraudulent or fraudulent behavior. These quantities we term p(x|NF) and p(x|F). Through postulating the possibility of fraud P(F) and P(NF) = 1P(F) in usual and through applying Bayes' rule, we acquire the fraud probability, provided the measurement x,

P(F|x) = P(F)p(x|F)/ p(x) where, the denominator p(x) could be calculated as P(x) = P(F)p(x|F) + P(NF)p(x|NF)

The chain rule of prospects is:

Assume there are 2 classes C_1, C_2 for fraud and legal correspondingly. Given an example

$X = (X_1, X_2, \ldots, X_n)$ and every row is represented through an attribute vector $A = (A_1, A_2, \ldots, A_n)$

The categorization is to derive the extreme $P(C_i|X)$ which could be derived from Bayes' theorem as mentioned in the following steps:

P(fraud|X) = [P(fraud | X) P(fraud)] / P(X)

P(legal|X) = [P(legal | X) P(legal)] / P(X)

As $P(X)$ is constant for entire classes, merely [*P*(*fraud* | *X*) *P*(*fraud*)] and [*P*(*legal* | *X*) *P*(*legal*)] need to be enhanced.

The class previous probabilities might be estimated through P(fraud) = s_i / s

Here, s is the entire number of training instances and s_i is the number of training instances of class *fraud.*

- A basic supposition of no dependence relation amongst attributes is formed.

Therefore,

$$P(X|fraud) = \prod_{k=1}^{n} P(x_k|fraud)$$

and

$$P(X|legal) = \prod_{k=1}^{n} P(x_k|legal)$$

The probabilities $P(x_1|fraud)$, $P(x_2|fraud)$ could be assessed from the training examples:

$P(x_k|fraud) = s_{ik} / s_i$

Here, s_i is the number of training instances for class *fraud* and s_{ik} is the number of training instances of class having value x_k for A_k

4.3.1.1. Application

We show a Bayesian learning algorithm to forecast the incidence of fraud. Utilizing the "Output" classification outcomes for Table 4.1, there are 17 tuples categorized as legal, and 3 as fraud. To assist categorization, we split the age of driver attribute into ranges:

Table 4.1. Training set instances

In-stance	Name	Gender	Age_ driver	fault	Driver_ rating	Vehicle_ age	Output
1	Kristina Green	F	28	1	0	4	legal
2	Michael Vasconi	M	36	0	0.33	4	legal
3	Crystal Smith	F	21	1	0.66	8	legal
4	Maggie Frazier	F	43	1	0.66	3	legal
5	Robert Howard	M	36	1	0.33	1	legal
6	David Okere	M	25	1	0	2	legal
7	Chibuike Penson	M	38	0	0.66	6	legal
8	Collin Pyle	M	41	1	0.33	3	legal
9	Eric Penson	M	38	1	1	2	fraud
10	Jeremy Dejean	M	41	0	0	7	legal
11	Beau Jackson	M	32	1	1	5	fraud
12	Jerry Smith	M	34	1	1	5	legal
13	Aaron Dusek	M	49	1	0.33	8	legal
14	Justin Howard	M	22	1	0	2	fraud
15	Derek Garrett	M	33	0	0	3	legal
16	Jasmine Jackson	F	28	0	1	2	legal
17	Chris Wilson	M	29	1	1	6	legal
18	Michael Pullen	M	41	1	0	5	legal
19	Bryan Thompson	M	33	1	0.33	4	legal
20	Bryan Sanders	M	48	1	0	3	legal
X	Crystal Smith	F	31	1	0	2	?

Table 4.2 displays the counts and following prospects related to the attributes. With these simulated training records, we approximate the preceding probabilities:

The classifier had to forecast the class of instance to be legal or fraud.

P(fraud) = s_i / s = 3/20 = 0.15

P(legal) = s_i / s =17/20 = 0.85

Table 4.2. Probabilities related to attributes

Attribute	Value	Count		Probabilities	
		fraud	legal	Fraud	legal
Gender	M	3	13	3/3	13/17
	F	0	4	0/3	4/17
age_driver	(20, 25)	0	3	0	3/18
	(25, 30)	0	4	0	4/18
	(30, 35)	1	3	1/2	3/18
	(35, 40)	1	3	1/2	3/18
	(40, 45)	0	3	0	3/18
	(45, 50)	0	2	0	2/18
fault	0	0	5	0	5/17
	1	3	12	3/17	12/17
driver_rating	0	1	6	1/3	6/17
	0.33	0	5	0	5/17
	0.66	0	3	0	3/17
	1	2	3	2/3	3/17

We utilize these values to categorize a novel tuple. Assume, we wish to categorize X = (Crystal Smith, F, 31). Through utilizing these values and the related probabilities of gender and driver age, we get the following estimates:

P(X |legal) = 4/17 * 3/18 = 0.039

P(X |fraud) = 3/3 * 1/2 = 0.500

Therefore, probability of being legal = 0.039 *0.9=0.0351

Probability of being fraud = 0.500 *0.1= 0.050

We approximate *P(X)* by summing up these individuals probability values subsequently *X* will be either legal or fraud:

P(X) = 0.0351 + 0.050 = 0.0851

Lastly, we get the real probabilities of every event:

P(legal | X) = (0.039 *0.9)/0.0851= 0.412

P(fraud |X) = (0.500 *0.1) / 0.0851= 0.588

Thus, founded on these probabilities, we categorize the novel tuple as fraud since its probability is highest.

As attributes are taken as autonomous, the adding of laid-off ones decreases its predictive strength. To reduce this conditional liberation is to add resultant attributes which are formed from groupings of prevailing attributes.

Lost records cause issues throughout the classification procedure. Naïve Bayesian classifier could handle lost values in training sets of data. To reveal this, 7 lost values seem in the dataset.

The naïve Bayes method is easy to utilize and only a single scan of the training records is needed. The method could handle lost values by simply neglecting that probability when computing the probabilities of membership in every class. Though the method is direct, it doesn't always provide satisfactory outcomes. The attributes generally are not autonomous. We could utilize a subset of the features by overlooking any that are relying on others. The method doesn't handle constant data. Splitting the constant values into ranges could be utilized to solve this issue, however, the division of the constant values is a boring task, and how this is completed could influence the outcomes.

Decision Tree-Based Algorithm

A DT (decision tree) is a tree related to a database that had every internal node categorized with an attribute, very arc categorized with a base that could be used to the attribute, and every leaf node considered with a class. Resolving the classification issue is a two-step procedure: i) decision tree induction construct a DT, and ii) implement the DT to decide its class. Rules could be produced that are easy to understand. They measure well for huge databases since the tree size is free of the size of the database.

DT algorithms don't simply handle constant data. The trait domains must be separated into categories. Handling lost is tough. Since the DT is created from the training data, overfitting might happen. This could be overcome through tree trimming (Chiu et al. 2004; Kim et al. 2002; Maes et al. 2002).

C4.5 Algorithm

The simple algorithm for the decision tree is as follows:

Assume there are 2 classes for legal and fraud The tree initiates as a particular node N demonstrating the samples of training.

If the examples are of a similar class fraud, then the node converts a leaf and is considered as fraud.

Then, the algorithm usages an entropy-founded measure as an experimental for choosing the feature that would best distinct the samples into single classes (Chen et al. 2004; Foster et al. 2004; Fan, 2004).

The expected or entropy information required to categorize a given sample is:

I(fraud, legal)= - (NumberFraudSamples / NumberSamples) log2 (NumberFraudSamples / NumberSamples) – (NumberLegalSamples / NumberSamples)

log2 (NumberLegalSamples / NumberOfSamples)

- Projected entropy or information needed to categorize into subsets through test attribute E is:

  E(A) = ∑ [(NumberTestAttributeFraudValues/ NumberSamples) +

  (NumberTestAttributeLegalValues/ NumberSamples)]*

  [I(TestAttributeFraudValues, TestAttributeLegalValues)]

- Probable decrease in entropy is:

Gain(A)= I – E(A)

The algorithm calculates the information gathered for every attribute. The attribute with maximum information gathering is the one chosen for the testing attribute.

- A branch is formed for every recognized value of the test feature. The algorithm utilizes the same procedure iteratively to create a decision tree for every portion. When an attribute had happened at a node, it required not be deliberated in any descendants of nodes.

The iterative segregating halts only when one of the situations is correct:

- all instances for a given node fit in the similar class, or

- there are no outstanding features on which trials might be further divided. If it is the situation, a leaf is formed with the majority class amongst samples,
- there are not any examples for the test attribute of the branch. In this situation, a leaf is formed with the main class in samples

4.3.2. Rule-Based Algorithm

One method to execute classification is to produce if-then rules. Some algorithms produce rules from trees also algorithms that produce rules without initial forming a decision tree (Syeda et al. 2002).

4.3.2.1. Producing Rules from a Decision Tree

The subsequent rules are produced for the Decision Tree.

If driver age =25, then class = legal

If (driver_age =40) ∧ (vehicle_age =7), then class = legal

If (driver_age =32)) ∧ (driver_rating =1), then class = fraud

If (driver_age ≤ 40)) ∧ (driver_rating =1)) ∧ (vehicle_age =2), then class = fraud

If (driver_age > 40)) ∧ (driver_age ≤ 50)) ∧ (driver_rating = 0.33), then class = legal

4.4. MODEL PERFORMANCE

4.4.1. Confusion Matrix

There are two methods to inspect the performance of the classifiers: i) confusion matrix, and ii) to utilize a ROC graph. Assumed a class, C_j, and a tuple, t_i, that tuple might or might not be allotted to that class however its real membership might or might not be in that class. Through two classes, there are 4 probable conclusions with the classification as i) false positives (false alarms), ii) true positives (hits), iii) false negatives, and iv) true negatives (accurate rejections). False-positive happens if the real outcome is legal however incorrectly foreseen as fraud. False-negative happens when the real outcome is fraud however incorrectly foreseen as legal. A confusion matrix (Provost and Kohavi, 1998), Table 4.3, comprises information regarding real and foreseen classifications. Performance is assessed by utilizing the records in the matrix. Table 4.4 displays the confusion matrix created on replicated data. It displays the classification model is being implemented to the trial data that comprises 7000 examples roughly divided evenly amongst two classes. The model makes some mistakes and had a precision of 78 percent. We also implement the model to the similar data, however to the negative class with regarding class tilt in the data. The quality model quality highly relies on the selection of the trial data. We also know that ROC curves are not so reliable on the selection of trial data, at least with class tilt (Stefano et al. 2001; Phua et al. 2004; Viaene et al. 2004).

Table 4.3. Confusion Matrix

legal		**Observed**	
		fraud	
Predicted	fraud	FN	TN
	legal	TP	FP

Table 4.4. Confusion matrix of a model implemented to the trial dataset

fraud		**Observed**		
		legal	accuracy: 0.78	
Predicted	legal	1125	3100	recall: 0.86
	fraud	2380	395	precision: 0.70

Several model performance metrics (Table 4.5) could be taken from the confusion matrix.

The *precision* determined in (Table 4.4) might not be a sufficient performance amount when the number of negative instances is much higher than the number of positive instances (Kubat et al.., 1998). Assume there are 1500 instances, 40 of which are positive instances and 1460 of which are negative instances. If the system categorizes them entirely as negative, the exactness would be 97.3 percent, however, the classifier lost all positive instances. Additional performance measures are *F-Measure* (Gale and Lewis, 1994) and *geometric mean* (*g-mean*) (Kubat et al.., 1998). For computing F-measure, β had a value from 0 to ∞ and is utilized to regulate the weight allotted to *P and TP*. Any classifier assessed utilizing *F measure or g-mean* would have a value of 0 if entire positive instances are categorized erroneously.

Table 4.5. Performance metrics

model performance metrics	
Accuracy(AC)	$AC = \frac{a+d}{a+b+c+d}$
False-negative rate(FN)	$FN = \frac{c}{c+d}$
True negative rate(TN)	$TN = \frac{a}{a+b}$
False-positive rate(FP)	$FP = \frac{b}{a+b}$
geometric mean(g-mean)	$g-mean_1 = \sqrt{TP*P}$ $g-mean_2 = \sqrt{TP*TN}$
F-measure	$F = \frac{(\beta^2+1)*P*TP}{\beta^2*P+TP}$
Precision(P)	$P = \frac{d}{b+d}$
Recall or true positive rate(TP)	$TP = \frac{d}{c+d}$

To effortlessly view and comprehend the output, visualization of the outcomes is supportive.

Naïve Bayesian visualization gives a collaborative view of the forecast results. The features could be sorted through the predictor and evidence matters could be sorted through the number of items in its storing bin. Trait column graphs assist to locate the important traits in neural networks. Decision tree visualization forms trees through splitting features from C4.5 classifiers (Brockett et al. 2002; Belhadji et al. 2000).

Lift charts and increasing gains are visual aids for computing model performance. Lift is an amount of a predictive model computed as the ratio amongst the results got without or with the predictive model. For example, if 105 of the entire samples are fraud and a naïve Bayesian classifier could appropriately predict 20 fraud trials per 100 trials, then that agrees to a lift of 4.

Table 4.6. Predictions Costs

fraud	legal
False Negative(miss) cost= number of misses* average cost per claim	True Negative(correct rejection) cost = number of correct rejection claims * average cost per claim
True Positive(Hit) cost= number of hits* average cost per investigation	False Positive(False alarm) cost=number of false alarms * (Average cost per investigation + average cost per claim)

Table 4.6 displays that False Positives (false alarms) and True Positives (hits) need a cost for every investigation. The false alarms costs are the most costly since both claim costs and investigation costs are needed. True Negatives(correct rejection) and False Negatives (misses) are the claim cost.

4.4.2. Relative Operating Characteristic Curve

An additional method to inspect the performance classifiers' performance is to utilize a ROC (Relative Operating Characteristic) curve, (Swets, 1988). A Relative Operating Characteristic graph is a curve that describes the performance tradeoff and performance of a classification model (Flach, 2004 Fawcett, 2004, Flach, 2004) through the False Positives alongside the X-axis and the True Positives alongside the *Y*-axis. The point (0, 1) is the ideal classifier: it categorizes all negative cases and positive cases properly. It is (0, 1) since the TP rate is 1 and the false-positive FP is 0. The point (0, 0) signifies a classifier that forecasts all instances to be negative, however, the point (1, 1) resembles a classifier that forecasts each instance to be positive. The point (1, 0) is the classifier that is inappropriate for entire classifications. A ROC point or curve is autonomous of error costs or class distribution (Provost et al.., 1998). It sums entire information comprised in the confusion matrix, as TN is the complement of FP and FN is the complement of TP (Swets, 1988). It gives a visual tool for inspecting the exchange amongst the number of negative instances erroneously classified and a classifier to properly identify positive instances.

We present a novel performance metrics to build ROC curves (in terms of confusion matrix), the FP Rate (FPR) and the TP Rate (TPR):

TPR(recall) = TP / (TP+FN)

FPR = FP / (TN +FP)

The classifier is charted to a similar point in the ROC graph irrespective of whether the actual trial set with tested down the negative class is utilized demonstrating that ROC graphs are not delicate to class skew (Moreau et al. 1997; Rosset et al. 1999).

One method of comparing ROC points is through utilizing an equation that associates precision with the Euclidian distance from the ideal classifier, the point (0, 1). We comprise a weight factor that permits defining relative costs of misclassification. AC_d is defined as a distance founded performance measure:

$$AC_d = 1 - \sqrt{W*(1-TP)^2 + (1-W)*FP^2},$$

where W ranges from 0 to 1, which is utilized to assign comparative significance to false negatives and false positives. AC_d extent from 0 for the ideal classifier to *sqrt(2)* for a classifier that categorizes entire cases erroneously. AC_d varies from *F-measure g-mean*$_1$, and *g-mean*$_2$ in that it is equivalent to 0 only if entire instances are classified properly. In additional words, a classifier assessed utilizing AC_d got some credit for precise classification of negative instances, irrespective of its precision in properly classifying positive instances (Cahill et al. 2002; Cortes et al. 2003).

REFERENCES

1. [uElkan, C. (2001). Magical Thinking in Data Mining: Lessons from CoIL Challenge 2000. *Proc. of SIGKDD01*, 426-431.
2. Barse, E., Kvarnstrom, H. & Jonsson, E. (2003). Synthesizing Test Data for Fraud Detection Systems. *Proc. of the 19th Annual Computer Security Applications Conference*, 384-395.
3. Belhadji, E., Dionne, G. & Tarkhani, F. (2000). A Model for the Detection of Insurance Fraud. *The Geneva Papers on Risk and Insurance* **25**(4): 517-538.
4. Bell, T. & Carcello, J. (2000). A Decision Aid for Assessing the Likelihood of Fraudulent Financial Reporting. *Auditing: A Journal of Practice and Theory* **10**(1): 271-309.
5. Bentley, P. (2000). Evolutionary, my dear Watson: Investigating Committeebased Evolution of Fuzzy Rules for the Detection of Suspicious Insurance Claims. *Proc. of GECCO2000.*
6. Bentley, P., Kim, J., Jung., G. & Choi, J. (2000). Fuzzy Darwinian Detection of Credit Card Fraud. *Proc. of 14th Annual Fall Symposium of the Korean Information Processing Society.*
7. Bhargava, B., Zhong, Y., & Lu, Y. (2003). Fraud Formalization and Detection. *Proc. of DaWaK2003*, 330-339.
8. Bolton, R. & Hand, D. (2001). Unsupervised Profiling Methods for Fraud Detection. *Credit Scoring and Credit Control VII.*
9. Bolton, R. & Hand, D. (2002). Statistical Fraud Detection: A Review (With Discussion). *Statistical Science* **17**(3): 235-255.
10. Bonchi, F., Giannotti, F., Mainetto, G., Pedreschi, D. (1999). A Classificationbased Methodology for Planning Auditing Strategies in Fraud Detection. *Proc. of SIGKDD99*, 175-184.
11. Brause, R., Langsdorf, T. & Hepp, M. (1999). Neural Data Mining for Credit Card Fraud Detection. *Proc. of 11th IEEE International Conference on Tools with Artificial Intelligence.*
12. Brockett, P., Derrig, R., Golden, L., Levine, A. & Alpert, M. (2002). Fraud Classification using Principal Component Analysis of RIDITs. *Journal of Risk and Insurance* **69**(3): 341-371.
13. Burge, P. & Shawe-Taylor, J. (2001). An Unsupervised Neural Network Approach to Profiling the Behavior of Mobile Phone Users for Use in

Fraud Detection. *Journal of Parallel and Distributed Computing* **61**: 915-925.

14. Cahill, M., Chen, F., Lambert, D., Pinheiro, J. & Sun, D. (2002). Detecting Fraud in the Real World. *Handbook of Massive Datasets* 911-930.
15. Chan, P., Fan, W., Prodromidis, A. & Stolfo, S. (1999). Distributed Data Mining in Credit Card Fraud Detection. *IEEE Intelligent Systems* **14**: 67-74.
16. Chen, R., Chiu, M., Huang, Y. & Chen, L. (2004). Detecting Credit Card Fraud by Using Questionnaire-Responded Transaction Model Based on Support Vector Machines. *Proc. of IDEAL2004*, 800-806.
17. Cortes, C., Pregibon, D. & Volinsky, C. (2003). Computational Methods for
18. Cox, E. (1995). A Fuzzy System for Detecting Anomalous Behaviors in Healthcare Provider Claims. In Goonatilake, S. & Treleaven, P. (eds.) *Intelligent Systems for Finance and Business*, 111-134. John Wiley.
19. Dynamic Graphs. *Journal of Computational and Graphical Statistics* **12**: 950970.
20. Ezawa, K. & Norton, S. (1996). Constructing Bayesian Networks to Predict Uncollectible Telecommunications Accounts. *IEEE Expert* October: 45-51.
21. Fan, W. (2004). Systematic Data Selection to Mine Concept- Drifting Data Streams. *Proc. of SIGKDD04*, 128-137.
22. Fawcett, T. (2004). ROC graphs: Notes and practical considerations for researchers. Machine Learning, 3.
23. Fawcett, T., & Flach, P. A. (2005). A response to web and Ting's on the application of ROC analysis to predict classification performance under varying class distributions. *Machine Learning,* 58(1): 33-38.
24. Flach, P. (2004). Tutorial at ICML 2004: The many faces of ROC analysis in machine learning. Unpublished manuscript.
25. Flach, P. A. (2003). The geometry of ROC space: Understanding machine learning metrics through ROC isometrics. *Proceedings of the Twentieth International Conference on Machine Learning,* 194–20 1.
26. Flach, P., Blockeel, H., Ferri, C., Hernandez-Orallo, J., & Struyf, J. (2003). Decision support for data mining: Introduction to ROC analysis and its applications. *Data mining and decision support: Aspects of integration and collaboration,* 81-90.

27. Foster, D. & Stine, R. (2004). Variable Selection in Data Mining: Building a Predictive Model for Bankruptcy. *Journal of American Statistical Association* **99**: 303-313.
28. He H, Wang J, Graco W and Hawkins S.(1997). Application of Neural Networks to Detection of Medical Fraud. Expert Systems with Applications, **13**, 329-336.
29. James F.(2002). FBI has eye on business databases. *Chicago Tribune*, Knight Ridder/ Tribune Business News.
30. Kim, H., Pang, S., Je, H., Kim, D. & Bang, S. (2003). Constructing Support Vector Machine Ensemble. *Pattern Recognition* **36**: 2757-2767.
31. Kim, J., Ong, A. & Overill, R. (2003). Design of an Artificial Immune System as a Novel Anomaly Detector for Combating Financial Fraud in Retail Sector. *Congress on Evolutionary Computation.*
32. Lin, J., Hwang, M. & Becker, J. (2003). A Fuzzy Neural Network for Assessing the Risk of Fraudulent Financial Reporting. *Managerial Auditing Journal* **18**(8): 657-665.
33. Little, B., Johnston, W., Lovell, A., Rejesus, R. & Steed, S. (2002). Collusion in the US Crop Insurance Program: Applied Data Mining. *Proc. of SIGKDD02*, 594-598.
34. Maes, S., Tuyls, K., Vanschoenwinkel, B. & Manderick, B. (2002). Credit Card Fraud Detection using Bayesian and Neural Networks. *Proc. of the 1st International NAISO Congress on Neuro Fuzzy Technologies.*
35. Magnify(2002). FraudFocus Advanced Fraud Detection, White Paper, Chicago.
36. Magnify(2002). The Evolution of insurance Fraud Detection: Lessons learnt from other industries, White Paper, Chicago.
37. Major, J. & Riedinger, D. (2002). EFD: A Hybrid Knowledge/ Statistical-based system for the Detection of Fraud. *Journal of Risk and Insurance* **69**(3): 309324.
38. McGibney, J. & Hearne, S. (2003). An Approach to Rules-based Fraud Management in Emerging Converged Networks. *Proc. Of IEI/IEEE ITSRS 2003.*
39. Meena J(2003). Data mining for Homeland Security. Executive Briefing, VA.

40. Meena J(2003). Investigative Data Mining for Security and Criminal Detection, Butterworth Heinemann, MA.

41. Moreau, Y. & Vandewalle, J. (1997). Detection of Mobile Phone Fraud Using Supervised Neural Networks: A First Prototype. *Proc. of 1997 International Conference on Artificial Neural Networks*, 1065-1070.

42. Ormerod T., Morley N., Ball L., Langley C., and Spenser C. (2003). 'Using Ethnography To Design a Mass Detection Tool (MDT) For The Early Discovery of Insurance Fraud', Computer Human Interaction, April 5-10, Ft. Lauderdale, Florida.

43. Phua, C., Alahakoon, D. & Lee, V. (2004). Minority Report in Fraud Detection: Classification of Skewed Data, *SIGKDD Explorations* **6**(1): 50-59.

44. Provost, F., Fawcett, T., & Kohavi, R. (1998). The case against accuracy estimation for comparing induction algorithms. *Proceedings of the Fifteenth InternationalConference on Machine Learning,* , 445–453.

45. Rosset, S., Murad, U., Neumann, E., Idan, Y. & Pinkas, G. (1999). Discovery of Fraud Rules for Telecommunications - Challenges and Solutions. *Proc. of SIGKDD99*, 409-413.

46. SAS e-Intelligence(2000). Data Mining in the Insurance industry: Solving Business problems using *SAS Enterprise Miner Software*, White Paper.

47. Shao, H., Zhao, H. & Chang, G. (2002). Applying Data Mining to Detect Fraud Behavior in Customs Declaration. *Proc. of 1st International Conference on Machine Learning and Cybernetics*, 1241-1244.

48. Sherman, E. (2002). Fighting Web Fraud. *Newsweek,* June 10.

49. SPSS(2003). Data mining and Crime analysis in the Richmond Police Department, White Paper, Virginia.

50. Stefano, B. & Gisella, F. (2001). Insurance Fraud Evaluation: A Fuzzy Expert System. *Proc. of IEEE International Fuzzy Systems Conference*, 1491-1494.

51. Swets, J. A., Dawes, R. M., & Monahan, J. (2000). Better decisions through science. *Scientific American; Scientific American, 283*(4), 82-87.

52. Syeda, M., Zhang, Y. & Pan, Y. (2002). Parallel Granular Neural Networks for Fast Credit Card Fraud Detection. *Proc. of the 2002 IEEE International Conference on Fuzzy Systems*.

53. Viaene, S., Derrig, R. & Dedene, G. (2004). A Case Study of Applying Boosting Naive Bayes to Claim Fraud Diagnosis. *IEEE Transactions on Knowledge and Data Engineering* **16**(5): 612-620

54. Von Altrock, C. (1997). Fuzzy Logic and Neurofuzzy Applications in Business and Finance. 286-294. Prentice Hall.

55. Weatherford, M.(2002). Mining for Fraud. *IEEE Intelligent Systems*, July/ August, 4-6.

56. Wheeler, R. & Aitken, S. (2000). Multiple Algorithms for Fraud Detection. *Knowledge-Based Systems* **13**(3): 93-99.

57. Williams, G. J. and Huang, Z.(1997). 'Mining the Knowledge: Mine the Hot Spots Methodology for Mining Large Real World Databases', 10th Australian Joint Conference on Artificial Intelligence, Published in Lecture Notes in Artificial Intelligence, Springer-Verlag, December, Perth, Western Australia.

58. Williams, G.(1999). 'Evolutionary Hot Spots Data Mining: An Architecture for Exploring for Interesting Discoveries', Proceedings of the 3rd Pacific-Asia Conference in Knowledge Discovery and Data Mining, Beijing, China

59. Yamanishi, K., Takeuchi, J., Williams, G. & Milne, P. (2004). On-Line Unsupervised Outlier Detection Using Finite Mixtures with Discounting Learning Algorithms. *Data Mining and Knowledge Discovery* **8**: 275-300.

Chapter 5

Application of Data Mining in Crime Detection

CONTENTS

5.1. INTRODUCTION

The extraction of information from huge data collections is a method called data mining. It is also the use of AI (artificial intelligence) techniques for the discovery of concealed relations between data. It is established as a significant area of study due to the data mining uses' extensive scope. For the application of data mining, criminology is among the most significant areas. The exploration and detection of crimes and the offender's connection with them are what constitute crime analysis. It is a suitable area for the application of data mining methods due to the data collections huge amount and the intricacy of the relations among them. For the development of analysis, the features of the crime need to be identified (Villafranca et al., 2012). The police can be helped and supported through the information attained from these methods which makes it a very valuable tool. The extraction of significant entities in simple text from police narrated accounts will be under discussion here through the use of data mining methods. In the organizations of law implementation, the database mechanically receives information about the crime through this approach. Crime matching techniques will be performed by using the clustering outcomes after applying the SOM clustering technique in crime analysis (Fei et al., 2005).

Encountering crime is inevitable in a human's lifetime. Looking at it as an effective combating skill, the know-how about crime analysis should be known by all. Crimes can be predicted, identified, and discovered by making use of a methodical approach which is what constitutes crime analysis. The crime factors are given knowledge and data which is the input of this structure while getting answers to the analytical and exploratory questions, extraction of information and finally, the outcome's visualization is the output (Lee & Huang, 2002). For crime detectives, crime analysts, and criminologists, this is a quickly spreading area due to the crime and the information related to its intricate nature and the concealed and imperceptible relations among them. The police organizations consist of information related to crime. However, the conventional techniques of crime analysis are no longer applicable due to the intricacy of relations among the data. Firstly, due to the factor of humans interfering, efficient relations/parameters are unable to be included and secondly, a large number of human resources and time are needed for these techniques. It has been concluded that for the investigation of crime, the usage of an intellectual and methodical approach is needed as revealed by such deficits. The chief solution can be the data mining methods (Agrawal et al., 2012).

5.2. FUNDAMENTALS OF INTELLIGENT CRIME ANALYSIS

In the procedure of crime analysis, 2 chief constituents are crime matching and crime variables. The analysis system primarily works on the crime variables which makes it very significant. Whereas, in intelligent crime uncovering, crime detection has a massive use. As basics of intelligent crime analysis, the following subjects are discussed below (Gupta et al., 2006).

5.2.1. Crime Variables

Crime features can be explained exceptionally by some parameters. In crime analysis procedure, crime variables are these exceptional crime parameters and are very significant to it. They can be sorted into 3 categories irrespective of the kind of crime:

- Criminal reports (such as criminal descriptions (race, sex, age, etc.)
- Crime natural conditions (such as the behavioral pattern of the criminal or features of the crime scene)
- Spatio-temporal crime variables (such as the period of the incident or the coordinates of the crime's location)

Particular crime variables are related to all crime kinds. For instance, theft's crime variables will not be similar to murder's crime variables (Keyvanpour et al., 2011).

For various kinds of larceny such as auto-theft, burglary, robbery, etc., there are different types of crime variables even in the range of the crimes of larceny. Hence, various types of crime variables need to be analyzed for various kinds of crimes. Listed in Table 5.1 are the most significant variables for BDH (Burglary from dwelling Houses).

Table 5.1. (BDH) Burglary Dwelling Houses' crime variables

Categories	Related variables
location of entry	Walls, roof, window, etc.
method of entry	Climbing, drilling, destroying, breaking, tunnel, etc.
type of dwelling house	Apartment, villa, bungalow, etc.
type of searching	tidy, untidy, all rooms, just one place, etc.
Location of exit	Walls, roof, window, etc.
methods of offender interaction with the environment	Lock the door after entering, manipulate the alarm, killing the watchdog, etc.

5.2.2. Crime Matching

Crime matching is the procedure of allocating offenders or crimes to the crimes that were previously unresolved or cracked. 2 features are included in crime matching in forensic inquiries (Taha & Yoo, 2015):

- The assumption is that one or more criminals who are responsible for a particular crime are arrested. The arrested criminals are allocated the previously unresolved crime through a crime matching procedure.
- A condition that a fresh unresolved crime is alerted to the police thus, the offenders responsible for it are not found yet. Hence, the technique of the act and the likely suspect are suggested on the basis of their accounts by crime matching (Wang et al., 2006).

Before their arrest, a large number of criminals commit various crimes. Detectives compile a list of the same kinds of crimes through running a few SQL inquiries on crime information that took place before to match when an offender is discovered to be accountable for a particular crime. There are 2 chief disadvantages of this conventional technique of crime matching: 1) the unlimited queries which have various WHERE clauses have to be tested as there is no universal decree for assigning WHERE stipulations in the SQL inquiry statement. 2) Precision is lacking due to the normal plainness of the queries. Three chief stages are included in the procedure of crime matching (Hornik & Kuan):

- Selection of features: In the analysis procedure, the crime characteristics are extracted which are thought to be efficient to be included. Crime variables have a subsection of features in the field of intelligent crime analysis (table 5.1). Effective techniques of feature selection are Forward, Genetic algorithm, and Best-First Selection (Oatley et al., 2004).
- Encoding: The features which are selected are encoded in a suitable way to accomplish the correct algorithms.
- Matching algorithm: The sets of criminals or crimes that are alike are revealed by the algorithms. The working of every algorithm is on the basis of its comparison gauges. As matching algorithms, various clustering algorithms, Tversky's Contrast Model, and KNN (K-Nearest Neighbor) can be used (Oatley et al., 2004).

Through binary encoding and ANN (Artificial Neural Network), the presentation of a crime matching method has been done.

5.3. COMPONENTS OF THE PLANNED TECHNIQUE: TOWARD A CRIME MATCHING OUTLINE

For the utilization of data mining methods in the area of intelligent crime analysis, the suggested technique has been presented. The fundamental constituents are 3 kinds of data mining methods used by a methodical approach: 1) For crime matching procedure, Neural Network as an engine, 2) crime information clustering and 3) as a division of text mining, Entity Extraction.

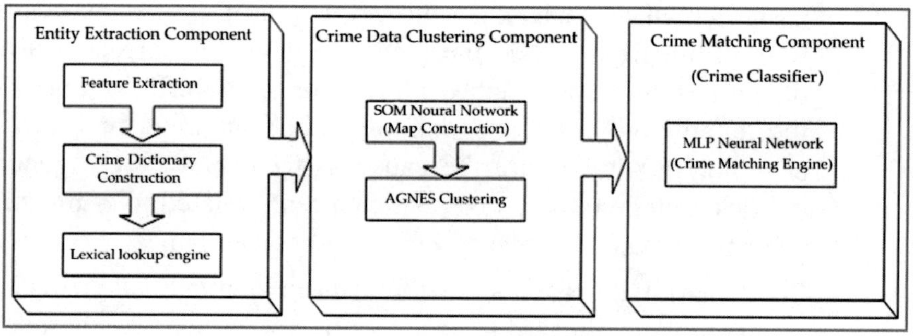

Figure 5.1. The components of the suggested technique

Source: https://www.semanticscholar.org/paper/Detecting-and-investigating-crime-by-means-of-data-Keyvanpour-Javideh/2ff18b62f19e3c1f7ed1562fa57be50ce18e499a

The discussion and dissection of these methods will be done. The relations among the constituents are displayed in Figure (5.1). The inner sub-constituents are also illustrated.

5.3.1. Mechanical Entity Withdrawal from Crime Account Reports: the Primary Constituent

There are four types of sets in which the general techniques for the extraction of entities from narrative accounts: 1) Rule-based, 2) Machine Learning-based techniques, 3) Lexical Lookup and 4) Static-based. In this study, the methodology of lexical lookup was utilized for crime entity removal. Selected by a crime field professional, the combined interested expressions and words were included in the lookup table created by the approach. Crime dictionary is what this lookup table is called. Amongst the textual information of police narrated accounts, the words from the dictionary were found by a plain search engine established by them. It is more difficult and time-consuming to manually read the narrative accounts and to insert them in an organized database while this method is more effective. By the leveraging of common words like "Mrs.", "Mr." and "organization" and more, the search engine's outcome is polished by the use of this approach (McGovern et al., 2007).

Outlined below are the difficult issues met in the procedure of entity extraction:

- The extraction of entities from different types of texts in the other issue domain is easier than the extraction of entities associated with crime from police narrated accounts. An extensive array of entities like coordinates of the site of the crime, model of the automobile, names of narcotics, and numbers included in criminology's range are the cause it is harder whereas general entities include mostly texts which are not related to crime such as names of establishments, names of people, names of places.
- Police narrated accounts are abundant in grammatical errors and misspelling. Compared to accounts like news text, it is harder to extract entities from information that is associated with crime due to these issues.

5.3.2. Crime Data Clustering

The discussion of the partition of clustering methods in intelligent crime analysis and leveraging classic hierarchal and the trials and deliberations associated with it is done. For the overcoming of various clustering methods disadvantages, the representation of a suggested approach is done which uses SOM neural network (Chen et al., 2003).

Grace is needed to commit analysis and clustering procedures on crime behavioral variables whose binary nature makes them out to be a trial. The infamous Euclidian distance measure – used for continuous kinds of variables- is rendered ineffective by binary encoding. In the clustering procedure, ambiguous outcomes can be achieved due to behaving binary measures such as continuous quantities (Keyvanpour et al., 2011). For the achievement of likeness amongst binary data materials, another kind of precise distance function should be exploited. The two items' dissimilarity is calculated as their paralleling distance by the functions. The following equation can calculate the dissimilarity (distance) between the 2 items (Oatley et al., 2004).

$D(i,j) = 1 - S(i,j)$ (1)

Amongst 2 binary arrangements *i* and *j*, D and S respectively stand for the functions of dissimilarity and similarity. Variables *a, b, c,* and *d* should be defined for the calculation of similarity as follows, supposing that both of the arrangements have the same length:

- In the 2nd bit arrangement, the number of bits that are equal to 0 is represented, while in the 1st arrangement, the ones equal to 1 are.
- In both bit arrangements, the representation is done of the number of respective bits with value 1.
- In both bit arrangements, the number of respective bits with the value 0 is represented. Lastly,
- In the 2nd bit arrangement, the number of bits that are equal to 1 are represented, while in the 1st arrangement, the ones equal to 0 are.

Table 5.2. Diverse methods of bit contrast

Respective variable	Bit value in the first sequence	Bit value in the second sequence
a	1	1
b	1	0
c	0	1
d	0	0

Through the equation below (Eqn. 2-4), the similarity amongst the bit arrangements ***i*** and ***j*** can be calculated based on the definitions above (Hand, 2007):

-	Simple Match Coefficient:	$s(i,j) = (p+s) / (p+q+r+s)$
-	Rao's Coefficient:	$s(i,j) = p / (p+q+r+s)$
-	Jaccard Coefficient:	$s(i,j) = p / (p+q+r)$

For clustering binary information items, the infamous standard ţ-means algorithm is unable to be utilized. The classic ţ-means' measured centroids are not binary is the cause. The middle of the cluster is represented by the binary item in the cluster in the ţ-medoids partitioning clustering technique which is proposed for the overcoming of the issue. The utilization of *centroid distance* or *Average-link* will be impractical for the tactic of inter-cluster distance measure if a hierarchical clustering technique is selected for the clustering of binary information items (Wrather et al., 2010). The distance measures *complete-link* or *single-link* can be employed (Wrather et al., 2010). With huge numbers of high-dimensional crime information, it is not beneficial to use hierarchical clustering because the time complexity of $O(n^2 logn)$ *and* $O(n^2)$, is found in various kinds of hierarchical clustering techniques (Oyang et al., 2001).

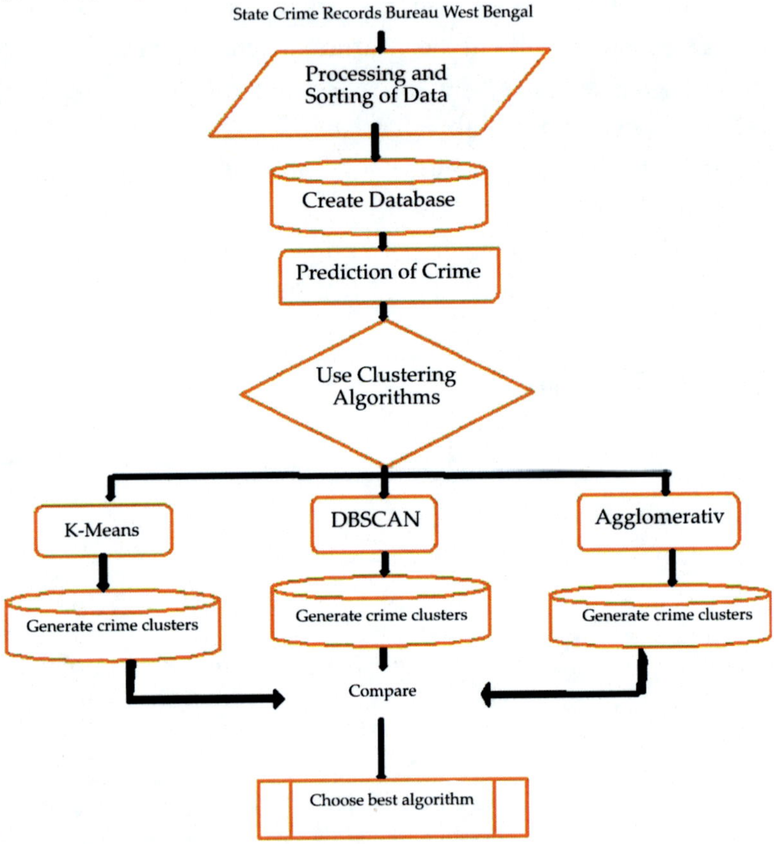

Figure 5.2. Crime frequency prediction's clustering method

Source: https://link.springer.com/chapter/10.1007/978-981-15-0035-0_35

5.3.3. Self-Organizing Map Neural Network: the 2nd Constituent

The tactic that may be selected for crime clustering may be influenced by the huge number of dimensions of information associated with crime. Multiple crime variables should be able to be dealt with by the approach. This issue can be dealt with by SOM's capability to plot the high-dimensional information spaces into low-dimensional views. The *data topology* is preserved whereas the number of dimensions is lessened by SOM neural network (Huysmans et al., 2006; Keyvanpour et al., 2011). The network is fed the statistical division of the system's output which is a low-dimensional manifestation of the high-dimensional information as to its input. For high-dimensional information picturing, this is seen as a graceful way. Not the topic of this section but *U-M*atrix and *Component Planes* are some beneficial conception algorithms (Keyvanpour et al., 2011; Ahmed et al., 2019). The natural inclination of SOM neural networks to be used in disseminated architectures and parallel processing is its most important benefit. Thus, a high volume of information associated with a crime can be efficiently dealt with by this technique. Also, information with non-linear statistical divisions can be dealt with by it (Keyvanpour et al., 2011). Hence, SOM capabilities are proposed to be exploited in this area because of its non-linear distribution of criminal information.

A 2-stage approach consisted of the suggested technique of crime information clustering; the feature map was extracted by the self-organizing neural system in the initial stage whereas, in the 2nd stage, the system output was sorted by the t-means infamous clustering algorithm. *However, what was the architecture of the neural system that was utilized?* By the procedure of feature extraction, binary numbers were attained which were the encoding of the twenty-one crime variables in the input layer. A 2D 25 x 25 range of neurons made up the output layer. Through weighted associations, every neuron in the input layer was connected to all neurons in the output layer. Through unplanned association weights, the initialization of the SOM neural system took place. The input layer received a few prearranged binary characters for the training of the system. Readjustment of the weights takes place in every repetition of the procedure. The actual high dimensional information division resembles the output layer by the time the training procedure finishes (Hurtado, 2004).

How does the procedure of training work? The resemblance is determined by the system amongst the output layer neurons and the input information by entering the system the train figures. A *winner neuron* is the one neuron in the output layer which resembles the input figure one. The computation of the space between the input information and the neurons of the output layer is done by the network receiving the input information. The *neighborhood function* value readjusts the neighbors of the winner and also its weight. Until the Θ (v) value is restricted to the winner neuron, it keeps lessening slowly throughout the procedure. The readjustment of the weights in a SOM neural network is shown in equation (5) (Hornik & Kuan, 1992).

$$Wv^{+} = Wv^{-} + \Theta(v) \times a \times (X - Wv^{-}) \qquad (5)$$

Where ɪ represents the *learning coefficient* whose value lessens throughout the training procedure, W_v^- is the neuron v's present weight, W_v^+ represents neuron *v's* modified weight that is allocated to the equivalent neuron, the input information is represented by X, and Θ (v) stands for the neighborhood function which was previously mentioned. The 2nd phase begins once SOM neural network produces the information map and the training stages end. The map produced by SOM goes through a hierarchical (like DIANA or AGNES) or a partitional (such as k-means) clustering algorithm. Sets like clusters are made of the SOM lattice (output layer) neurons as they are nearest to one other in weight. More analysis is done through utilizing them like crime matching which will be examined later. A hierarchical bottom-up clustering procedure i.e., AGNES (8) is employed for the 2nd phase. The precise yet easy technique is the reason AGNES was selected. However, space intricacy (*O(n3)*) and high time plague AGNES. A low-time intricacy is owned by the k-means algorithm which can be utilized to lessen this issue (Hurtado, 2004). In the k-means algorithm, unexpected initialization can be avoided through this opportunity in the estimation of the perfect *initialization seeds* and *cluster numbers* by the exploitation of the graphical map (SOM's output).

5.3.4. Suggested Crime Matching engine: Crime Classifier Constituent

For the analysis and investigation of crimes, one of the most significant needs in crime matching. Its goal is to match criminals to crime and vice versa. For crime matching, there are two practical methods as mentioned previously (Zeng et al., 2018):

- The aim is to find likely criminals who are responsible for a freshly occurred crime, and
- Previously unsolved cases are allocated to the identified criminal

The chief focus though is to use behavioral burglary crime variables with crime matching. Four sets were made to classify the burglary crime variables:

- Kinds of burglary places,
- The interaction of the criminal with the environment of crime

- Entry technique kinds
- The usage of tools and gadgets by the criminal. Binary encoded for entry technique kind is a list of crime variables in the table (5.3).

Table 5.3. System for entry techniques in a particular crime instance

Entry Methods	Binary encoded value
Front Door	0
Rear Door	0
Window	1
Breaking the Window	1
Climb	1
Damage Locks	0
Terrace	1

What is the working of the crime matching engine? Having a *back-propagation* training technique, the MLP (Multi-Layer Perceptron) neural system is utilized for a *classification* procedure (Sulaiman et al., 2009). Parallel information processing, as well as the toleration of loud information circumstances, are 1 of the most important advantages of using an MLP classifier. For entry technique kinds, kinds of the interaction of the criminal with the environment of crime, the usage of various kinds of tools and gadgets by the criminal and kinds of burglary places, and all sections of burglary crime variables are assigned an MLP neural network. Output, hidden and the input layer make up the MLP's network topology i.e., it has three layers. The approximation of any non-linear or linear function can be done if the architecture of the neural network contains even a single hidden layer (Honik, 1991). The number of crime variables in the analysis should be equivalent to the input layer's neurons. For instance, the input layer should have 8 neurons in the planned neural system of kinds of entry techniques (table 5.3 shows a single neuron for each feature). As established by the crime information clustering constituent, the number of clusters should be equivalent to the neurons in the output layer. The perfect number of neurons in the hidden layer can not be ascertained through a typical law, but it is proved that if the neurons of the hidden layer are selected as per the equation, then the MLP system will be more effective (Hauck et al., 2002).

$$m = a \times \sqrt{n_p \times n_a} \qquad (6)$$

Where m is the suggested number of hidden layer neurons, n_p and n_a represent the number of input and output layer neurons respectively and finally *a* is a coefficient that was set to be 4 based on some try-and-error experiments.

In which *a* represents a coefficient that had to be four on the basis of a few try-and-error trials, n_a and n_p stand for the number of output and input layer neurons correspondingly, and lastly, m is the proposed number of neurons in the hidden layer.

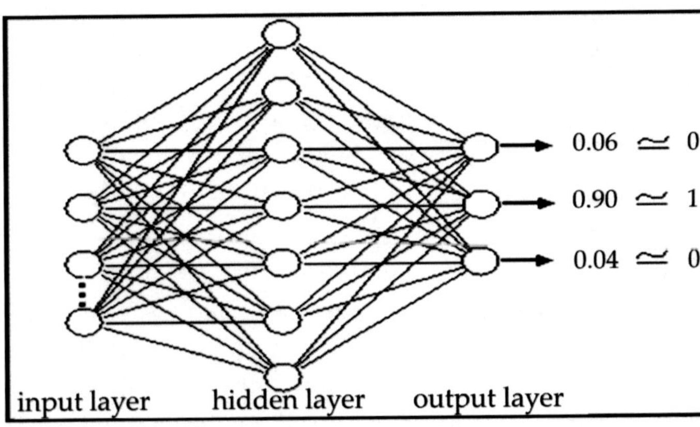

Figure 5.3. Interpretation of the output of MLP

Source:https://www.sciencedirect.com/science/article/pii/S1877050910005181

REFERENCES

1. Agrawal, T., Gupta, G. K., & Agrawal, D. K. (2012). Vitamin D deficiency decreases the expression of VDR and prohibitin in the lungs of mice with allergic airway inflammation. Experimental and molecular pathology, 93(1), 74-81.
2. Ahmed, R. F. M., Salama, C., & Mahdi, H. (2019). Optimizing Self-Organizing Maps Parameters Using Genetic Algorithm: A Simple Case Study. In International Conference on Advanced Intelligent Systems and Informatics, 3(2), 3-12.
3. Chauhan, R., Kaur, H., & Alam, M. A. (2010). Data clustering method for discovering clusters in spatial cancer databases. International Journal of Computer Applications, 10(6), 9-14.
4. Chen, H., Atabakhsh, H., Petersen, T., Schroeder, J., Buetow, T., Chaboya, L., ... & Huang, Z. (2003). COPLINK: Visualization for crime analysis. In Proceedings of the 2003 annual national conference on Digital government research, 2, 1-6.
5. Chen, H., Chung, W., Xu, J. J., Wang, G., Qin, Y., & Chau, M. (2004). Crime data mining: a general framework and some examples. computer, 37(4), 50-56.
6. Fei, B., Eloff, J., Venter, H., & Olivier, M. (2005). Exploring data generated by computer forensic tools with self-organising maps. Proceedings of the IFIP Working Group 11.9 on Digital Forensics, 1-15.
7. Gouda, G. R., Rao, M. S., & Soni, A. (2009). Web of Trust: An approach towards Semantic based Social Networks. International Journal of Recent Trends in Engineering, 1(1), 450.
8. Guo, P., Tang, R., Cheng, C., Xi, F., & Liu, M. (2005). Interfacial Organization-Induced Supramolecular Chirality of the Langmuir–Schaefer Films of a Series of PPV Derivatives. Macromolecules, 38(11), 4874-4879.
9. Gupta, G. K., Hecht, E. S., Zhu, H., Dean, A. M., & Kee, R. J. (2006). Gas-phase reactions of methane and natural-gas with air and steam in non-catalytic regions of a solid-oxide fuel cell. Journal of Power Sources, 156(2), 434-447.
10. Hand, D. J. (2007). Principles of data mining. Drug safety, 30(7), 621-622.

11. Hauck, R. V., Atabakhsb, H., Ongvasith, P., Gupta, H., & Chen, H. (2002). Using Coplink to analyze criminal-justice data. Computer, 35(3), 30-37.
12. Honik, K. (1991). Approximation capabilities of multilayer feedforward network. Neural Networks, 4(2), 251-257.
13. Hornik, K., & Kuan, C. M. (1992). Convergence analysis of local feature extraction algorithms. Neural Networks, 5(2), 229-240.
14. Hurtado, J. E. (2004). An examination of methods for approximating implicit limit state functions from the viewpoint of statistical learning theory. Structural Safety, 26(3), 271-293.
15. Huysmans, J., Martens, D., Baesens, B., Vanthienen, J., & Van Gestel, T. (2006). Country corruption analysis with self organizing maps and support vector machines. In International Workshop on Intelligence and Security Informatics, 3(1), 103-114.
16. Jiji, G. W., & Ananth: radha, S. (2012). Automatic tracking of criminals using data mining techniques. Journal of The Institution of Engineers (India): Series B, 93(4), 217-221.
17. Keyvanpour, M. R., Javideh, M., & Ebrahimi, M. R. (2011). Detecting and investigating crime by means of data mining: a general crime matching framework. Procedia Computer Science, 3, 872-880.
18. Kwok, K. O., Li, K. K., Wei, W. I., Tang, A., Wong, S. Y. S., & Lee, S. S. (2021). Influenza vaccine uptake, COVID-19 vaccination intention and vaccine hesitancy among nurses: A survey. International journal of nursing studies, 114, 103854.
19. Lee, S. C., & Huang, M. J. (2002). Applying AI technology and rough set theory for mining association rules to support crime management and fire-fighting resources allocation. Journal of Information, Technology and Society, 2(65), 65-78.
20. Mande, U., Srinivas, Y., & Murthy, J. (2012). Criminal identification system based on facial recognition using generalized gaussian mixture model. Asian J. Comput. Sci. Inf. Technol, 6, 176-179.
21. Mande, U., Srinivas, Y., Murthy, J. V. R., & Kakinada, V. V. (2012). Feature specific criminal mapping using data mining techniques and generalized gaussian mixture model. Int J Comput Sci Commun Netw, 2(3), 375-379.

22. McGovern, T. H., Vésteinsson, O., Friđriksson, A., Church, M., Lawson, I., Simpson, I. A., ... & Dunbar, E. (2007). Landscapes of settlement in northern Iceland: Historical ecology of human impact and climate fluctuation on the millennial scale. American anthropologist, 109(1), 27-51.
23. Oatley, G. C., Zeleznikow, J., & Ewart, B. W. (2004). Matching and predicting crimes. In International Conference on Innovative Techniques and Applications of Artificial Intelligence, 2 (1), 19-32.
24. Oyang, Y. J., Chen, C. Y., & Yang, T. W. (2001). A study on the hierarchical data clustering algorithm based on gravity theory. In European Conference on Principles of Data Mining and Knowledge Discovery, 6(2), 350-361.
25. Paul, S., MacLennan, J., Tang, Z., & Oveson, S. (2005). Data Mining Tutorial. Microsoft Corporation, 50-59.
26. Peek, N., & Swift, S. (2012). Intelligent Data Analysis for Knowl edge Discovery, Patient Monitoring and Quality Assessment. Methods of information in medicine, 51(04), 318-322.
27. Sulaiman, S. I., Musirin, I., & Rahman, T. K. A. (2009). Prediction of grid-photovoltaic system output using three-variate ANN models. WSEAS Transactions on Information Science and Applications, 6(8), 1339-1348.
28. Sylvius, N., Bilinska, Z. T., Veinot, J. P., Fidzianska, A., Bolongo, P. M., Poon, S., ... & Tesson, F. (2005). In vivo and in vitro examination of the functional significances of novel lamin gene mutations in heart failure patients. Journal of medical genetics, 42(8), 639-647.
29. Taha, K., & Yoo, P. D. (2015). SIIMCO: A forensic investigation tool for identifying the influential members of a criminal organization. IEEE Transactions on Information Forensics and Security, 11(4), 811-822.
30. Vapnik, V. N. (1999). An overview of statistical learning theory. IEEE transactions on neural networks, 10(5), 988-999.
31. Villafranca, M. H., Bustillo, C. W. G., Cárdenas, V. T., & Pérez, Y. C. (2012). Escalamiento Multidimensional y Mapas Autoorganizados para visualizar el uso de los Métodos Estadísticos no paramétricos en la rama de las Ciencias Agraria y Biológica. Ciencias de la Información, 43(1), 51-56.

32. Wang, G. A., Chen, H., Xu, J. J., & Atabakhsh, H. (2006). Automatically detecting criminal identity deception: an adaptive detection algorithm. IEEE Transactions on Systems, Man, and Cybernetics-Part A: Systems and Humans, 36(5), 988-999.
33. Wrather, A., Shannon, G., Balardin, R., Carregal, L., Escobar, R., Gupta, G. K., ... & Tenuta, A. (2010). Effect of diseases on soybean yield in the top eight producing countries in 2006. Plant Health Progress, 11(1), 29.
34. Zeng, L., Tang, Z., Chen, P., Hou, C., & Chen, G. (2018). Bioacoustic application on fisheries management in an artificial reefs' ecological reserve of Bohai Gulf China. Environmental Earth Sciences, 77(21), 1-11.

Chapter 6

Data Mining in Telecommunication Industry

CONTENTS

6.1. INTRODUCTION

The data mining applications in telecommunications industry, and a learning system for decision support in telecommunications case study, knowledge processing in control systems, and aircraft control case study are discussed in this chapter. A few scenarios where data mining may improve telecommunication services are discussed (Joseph, 2013).

The deregulation of the telecommunications industry in many countries and the development of new computer and communication technologies and the telecommunication market are rapidly expanding and highly competitive. This creates a great demand for data mining in order to help understand the business involved, identify telecommunication patterns, catch fraudulent activities, make better use of resources, and improve the quality of service (Nadaf & Kadam, 2013).

In order to determine the needs of the telecommunication industry with respect to the data mining, an extensive literature survey analysis was performed at Telkom. They expressed a need for mining the data stored in the Telkom data warehouse. Almost all areas of Telkom's business can benefit from data mining, but in particular marketing and sales department (Keramati et al., 2014). A serious problem for Telkcom, and for most companies in the telecommunications industry, is the *problem of churning*. Churning is the process of customer turnover.

A case study for decision support in telecommunications has been described. History data describing the operation of a telephone exchange is analyzed by the system to reconstruct understandable event descriptions (Eze et al., 2017). This case study is taken from Gerstner Laboratory, Czech Technical University, Czech Republic.

Real-time knowledge-based or knowledge-processing systems are playing an increasingly important role in transportation, manufacturing, control, and robotic and aerospace systems. They are no longer limited to low-level control functions. Control, supervision, and monitoring of complex hierarchical systems in dynamic and sometimes unpredictable or hazardous environments are typical tasks of current man-made systems (Folasade, 2011).

Current development in real-time artificial intelligence is driven by a need to make knowledge-based systems work in real-time and a need to integrate knowledgebased approaches to handle the complexities of problem-solving behavior in control systems. The purpose is to present a real-time knowledge processing (RTKP) procedure based on conjunctive and disjunctive matrices and operators (Coussement et al., 2017).

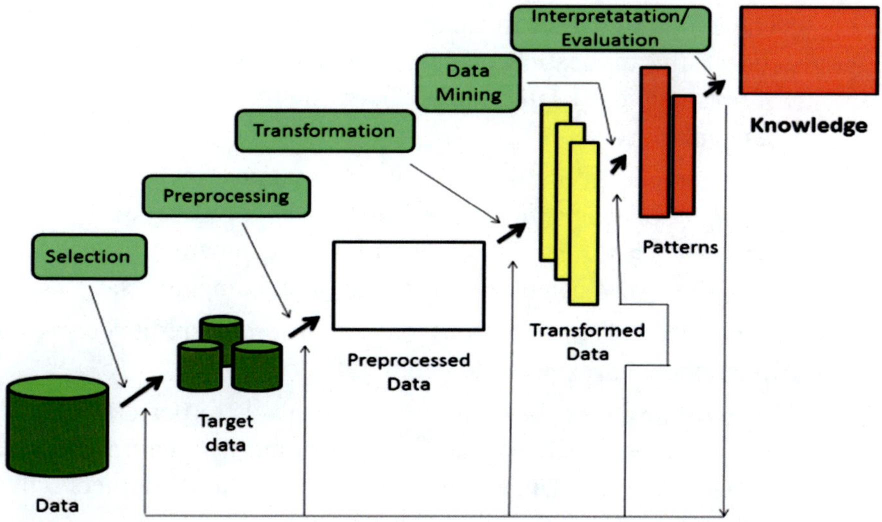

Figure 6.1. Data mining application process in telecommunication industry

Source: https://ijritcc.org/index.php/ijritcc/article/view/5218

The case study taken from Sylvain Letourneau, University of Ottawa, Canada, is discussed to explain the how data mining is used for maintenance of complex systems. The anticipated contributions of this study were related to two fundamental problems in the field of knowledge discovery in databases: i) automatic preparation of the data prior to model development and ii) use of diverse sources of information (Amin et al., 2017).

6.2. ROLE OF DATA MINING IN TELECOMMUNICATION INDUSTRY

- Data mining in telecommunication industry helps to understand the business involved, identify telecommunication patterns, catch fraudulent activities, make better use of resources, and improve the quality of service (Turhan et al., 2009).
- A large class of data mining algorithms developed for this purpose includes CART, C4.5, neural networks, and Bayesian classifiers, among others. One of the assumptions made by these algorithms,

which are carried over into data mining applications, is that of clean data.

- The ability to handle noise in this case is obviously critical to the successful application of data mining algorithms; the treatment of noise typically falls short of handling the complete problem of data error.
- The companies in the telecommunications industry face the *problem of churning*. Churning is the process of customer turnover. This is a major concern for the companies having many customers who can easily switch to other competitors.
- Data mining is one solution to do appropriate credit scoring and to combat churns in the telecom industry.
- Data mining may be used in churn analysis to perform two key tasks: Predict whether a particular customer will churn and when it will happen; Understand why particular customers churn (Ewieda et al., 2021).
- Decision support in telecommunications forms the rules that can be used as decision support rules (for the exchange operator) or directly to automate the operation of the exchange.
- In control systems the purpose is to present a real-time knowledge processing (RTKP) procedure based on conjunctive and disjunctive matrices and operators.
- The field of knowledge discovery in databases (KDD) has delivered a variety of techniques to discover patterns from vast amount of data, which helps in mining for complex data (Ramageri & Desai, 2013).

6.3. DATA MINING AND TELECOMMUNICATION INDUSTRY

The telecommunication industry has quickly evolved from offering local and long-distance telephone services to providing many other comprehensive communication services, including voice, fax, pager, cellular phone, images, e-mail, computer, and Web data transmission, and data traffic. The integration of telecommunication, computing network, Internet, and numerous other means of communications and computing is also underway (Binti et al., 2010).

The following are a few scenarios where data mining may improve telecommunication services.

6.3.1. Multidimensional Analysis of Telecommunication Data

Telecommunication data are intrinsically with dimensions such as calling time, duration, location of caller, and type of call. The multidimensional analysis of such data can be used to identify and compare the data traffic, system work load, resource usage, user group behavior, profit, and so on. For example, an analyst in the industry may wish to regularly view charts regarding calling source, destination, volume, and time-of-day usage patterns. Therefore, it is often useful to consolidate telecommunication data into large data warehouse and routinely perform multidimensional analysis using OLAP and visualization tools (Gheware et al., 2014).

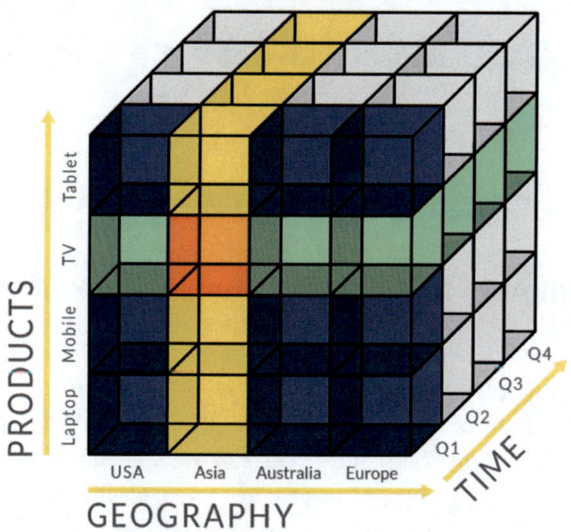

Figure 6.2. Online Analytical Processing (OLAP) analysis chart

Source: https://olap.com/olap-definition/

6.3.2. Fraudulent Pattern Analysis and the Identification of Unusual Patterns

Fraudulent activity costs the telecommunication industry millions of dollars a year. It is important to identify potentially fraudulent users and their atypical usage patterns; detect attempts to gain fraudulent entry to customer accounts; and discover unusual patterns that may need special attention, such as busy-hour, frustrated call attempts, switch and route congestion patterns, and periodic calls from automatic dial-out equipment (like fax machine) that have been improperly programmed (Ngai et al., 2011). Many of these types of patterns can be discovered by multidimensional analysis, cluster analysis, and outlier analysis (Leite et al., 2017).

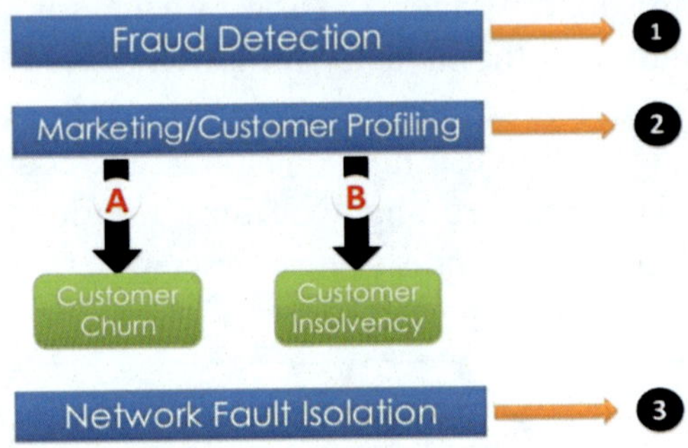

Figure 6.3. Fraud pattern detection and analysis

Source: https://www.slideshare.net/nabil_alsharafi/data-mining-140063251

6.3.3. Multidimensional Association and Sequential Pattern Analysis

The services of telecommunication can be enhanced in the multidimensional examination by the detection of sequential and association arrangements. If for instance, some consumers required services of communication by the day time and by month, and there is a need to search for usage arrangements, the consumer may group the records of the calling in pattern given below (Ko et al., 2016):

(Customer id, residence, office, time, date, service 1, service 2,...)

An arrangement of the sequence, "a consumer living in New York travles to work in some other state from his residence, there are chances that the cosumer would utilize services of long distance between the two states at 4 pm and for at least 4t minutes he would utilize cellular phone in the following hour in every working day" may be then explored to calculate its working for specific states and for specific persons by drilling up and down (Kirkland et al., 1999; Gray & Debreceny, 2014). This would intensify the sales of the particular cellular phone and long-distance groupings, and the accessibility of the specific services in an area are enhanced.

6.3.4. Use of Visualization Tools in Telecommunication Data Analysis

Apparatuses for linkage visualization, clustering association visualization, OLAP visualization, outlier visualization etc. are presented to be valuable for data investigation of communication (Manunza et al., 2017).

6.4. DATA MINING FOCUS AREAS IN TELECOMMUNICATION

Bell Atlantic STC, New York performed a grouping of the business of telecommunication with their applications of concern, and it was focused on the algorithms of the machine learning. Developers and researchers of the algorithms of the machine learning have establisehed and developed a class of aalgorithems of data mining from the concepts which have already been studied by the early researches (Bose & Mahapatra, 2001).Some of the noteworthy algorithms are neural networks, CART, Bayesian classifiers, and C4.5. Clean data is one of the main supposition which these algorithems make, that are performed into the applications of the data mining. These algorithms along with other of these kinds may also ease the postulation from its firmest standings. Perfectly clean data is not assumed by these algorithms, however, there is an assumption that data might be noisy. The treatment and handling of the noise is usually not up the mark as compared to the treatment of the whole issue of the error in data, while the handing and treatment of the noise is considered a crucial step towards the success of the application in the algorithms of the data mining (Hung et al., 2006; Jadhav & Pawar, 2011).

6.4.1. Systematic Error

There are some reasons given below due to which there are chances of the systematic errors in various applications of the data mining (Sumathi & Sivanandam, 2006):

- Human errors: errors due to the behaviors of different observes
- Issue in instrument calibration.
- Defective method.

In various applications of the telecommunications, there can be multiple examples which have been studied by Bell Atlantic STC, New York. Grouping of the consumer reported issues of telephone is one of the application in the telephone network having a local loop. Diagnoses of the issues are of high level, which roughly defines the portions of the local loop in which there aare chances of the trouble, hence there is a need to send a suitable specialist to resolve the issue (Toor et al., 2020). Dispatching of these diagnoses is carried out to the premise of the customer: communication to the cable, communication to the central office, keep for the additional testing. The data which has been acquired from the issue tells information abou the kind of switch used to which the line of the customer is connected and the further electrical measurements inculuding resistance, voltage etc. The consideration of the previous databse of the issues and ways in which those were resolved is an important step to resolve the data mining issue in hand, and to establish guidelines for dispatching the suitable experts out to resolve issue that have a definite summary (Dutt et al., 2017).

An automatic system of line testing is used to collect data of the electrical measurements which include data of in bulk quantities. There is a need that the calibration of the line testing must be performed regularly, but this seldom happens in practice. The system becomes miscalibrated as a consequence, and a systematic error may arise in all the measurments which are being reported on a certain dat for a set of lines. Morevover, day to day variation on the baseline measurement of the system may also occur (Shadroo & Rahmani, 2018).

The origin of the type of the systematic error is understood, however, to eliminate this error from the data, no mechanism is yet in practice. The cautious calibration is not given a high priority in a situation where company is handing a heavy load. Hence, the chances that the issue will remain are anticipated. Human errors are also possible which may also induce systematic error in the data. Specifically, one cause of the analyses for issue is the expert who resolve the issues. Colicated coding mechanisms are used to report the problem by the experts (Li & Beaubouef, 2010). There are chances that expert memorizes a wrong code which is reported to the central office then the issue will remain regularly. Hence, there are bright chances to reach the origin of the problem, however, mechanism to control these problems are still not clear. Moreover, system for the automated correction of these errors is not in practice, aside from aside from keeping a profile of each expert.

Regaring a ststematic error in data, there are various aspects which need to be considred (Umayaparvathi & Iyakutti, 2012; Balasubramanian & Selvarani, 2014).

- In scenarios where the systematic error is well known. In such situations, claning of the data can be performed by the algorithms of the data mining.
- The issue may be resolved. In some applications, data can be collected from a number of sources. In such scenarios, there is a possibility of data retention which are unchanging over the sources. By making a supposition that there are errors in the data and these errors are chaning with the number of sources, there will be influence on the cleaning of the data. To resolve the analytical error, a numbr fof data sources were utilized with the application local-loop diagnosis (however calibration errors were not taken into account).
- Cleaning of data may not be possible. In some scenarios, there is error present but it cannot be eliminated from the data. In such scenrios, it is important that the origin of these errors should be identified, however, due to these further difficulties it is hard to eliminate the error from the data.

An evident response to these scenarios is to let these cases go and accept that the application of methods of the data mining would give valuable outcomes. Hoewver, this response is not satisfactory (Kraljević & Gotovac, 2010; Shaaban et al., 2012).

- If the appropriate alogorithm is utilized, or the extent of the systematic error is not too great, the effects of the errors comparative to the benefits of the data mining are neglibible.
- The methods of the data mining could be beneficial for facilitating to recognize systematic error, crating a possibility of the data cleaning.
- In some applications, mined information of very little value can be beneficial for the company. In such scenarios, as data miners it is not feasible to just discard the application labelling it as "too hard". In the application stated above, an upgrading of only 1% over the present communication process might protect the company over $3,000,000 yearly.

Ther is a need to do improve:

- Development of the algorithms of the data mining for the elimination of the systematic error
- Examining the tools which have been used to measure the extent to which these have been affected by various kinds of errors (Hashmi et al., 2013).

6.4.2 Data Mining in Churn Analysis

Extraction of knowledge from the data is known as data mining. It includes usage of a number of tools from neural networks to the classical statistical approaches and other novel methods coming from artificial intelligence and machine learning. Lately, the use of data mining has enhanced the performance by refining the process optimization database advertising, and identifying fraud (Almana et al., 2014).

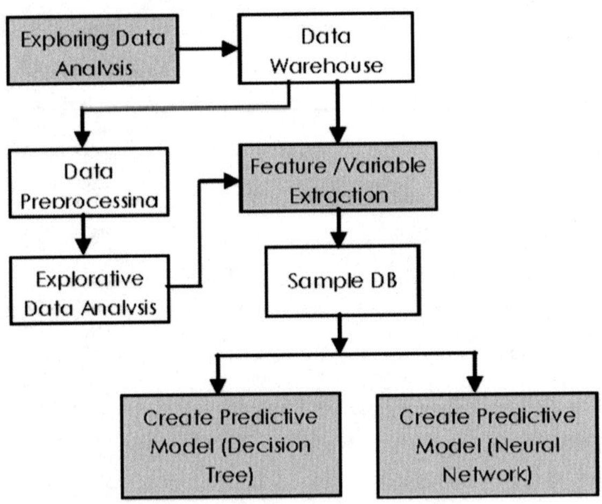

Figure 6.4. Churn prediction model

Source: https://www.semanticscholar.org/paper/Applications-of-Data-Mining-Techniques-in-Telecom-Umayaparvathi-Iyakutti/4f96e06db144823d16516af787e96d13073b4316/figure/0

Data mining can be valuable to every area of the Telkom's business, particularly sales and marketing sectors. The issue of the churning is the main issue for Telkcom and other telecommunication companies (Nath & Behara, 2003). Customer turnover can define the term churning. Companies having a large number of consumers can face this issue when consumer can shift to another competitor network. With the emergence of new companies in South Asia, the competition will become even severe. There would be rise in the churn rate with the rise in the choice of the consumer. In the cellular phones the churn rate is anticipated to rise 30% per annum (Mahajan et al., 2015).

Digital Equipment Corporation published a report in 1995 which anticipated the price of the churning in the communication of the wireless to be about $400 per new subscriber. It is obvious fact that costs on holding on to present consumer is more effective than obtaining the new consumer.

There are two main tasks to be performed in data mining to be used in churn examination:

Anticipate if a specific cosumer will churn and will occur; Recognize why specific consumer churn.

The recognition of tasks and anticipation show the two most significant features of the data mining in usage now a days. The chances of churn can be decreased by anticipating which consumers are probable to churn by proposing consumers new incentives to stay (Wei & Chiu, 2002). By anticipating which consumers are probable to churn the company may also work on shifting their service so as to please these consumers pro-actively. Along with the data mining tools so as to select the suitable approach in terms of effort and price can measure the chance of the consumer churning after act is engaged.

6.5. A LEARNING SYSTEM FOR DECISION SUPPORT IN TELECOMMUNICATIONS – CASE STUDY

In telecommunications, a mechanism of decision support is presented. The operation and working of the telephone exchange is examined by the system to establish the comprehensible event explanations as described by the history of the data. An algorithm processes the descripton of the event bringing rules explaining uniformities in the events. These rules may be utilized as decision support rules or openly to systematize exchange operation (Daskalaki et al., 2003).

The operators have no direct product of their own, however, they have powerful influence on the efficiency of the other workers. Inspite of this fact, there is a huge number of companies to offer the telephone operator post. The fact behind it is that it is difficult to find a person who is intelligent enough to perform operator tasks and modest enough to be just an operator

(Berkani, 2021). Then computers come in and take the job of the operator, computers have only capital investment in their costs. Additionally, there is non stop working of the machines and these can also provide extra data appropriate for the examination permitting for developments of the telecommunication traffic. Presently, a number of domains have involved computers in the area of PBX (ignoring the point that PBX itself is a type of computer) (Coussement et al., 2017; Yang et al., 2018):

- *Automated attendant* – a gadget used to to welcome a caller in an integrated way and permits him normally to reach some people, or give choice of persons from the spoken list; in anyway the caller is needed to co-operate.
- *Voice mail* – a gadget letting to leave a oral message to an inaccessible person and few rather formal ways of transporting the messages are existing.
- *Information service* –a person replacing device in giving few basic information generally arranged into a tree of information; the calling party is needed to cooperate.

The objective of the devices mentioned above is to provide convenience to a caller even when there is no service of human offering at the time. But all that kind of gadgets always act the same manner as they are created in a easy, static way. The purpose is easy – as, against the human operator, they do not show concern with who is calling nor what they generally need. We can consider the following the modification of the automated telephony by Comparing a compter to a human operator/receptionist (Shamsudin et al., 2019; Liu et al., 2020):

- The system can learn to find the most possibly wanted person by the caller by observing what number was dialed by the caller and who is calling (recognized by the calling party number); information can be attained either by "considering" the way how humans handled the caller earlier or from preceding cases (considering other information like explicit information ,daytime,– long absenteeism of few of the workers of a company, etc.); this could help the caller to access the information she desires.
- A machine, in a verbal language, can tell the caller about the condition of the call and recommended most suitable substitutions; messages should be "context sensitive."

Normally, the aim of the computerized telephone system is to integrate the functions of speaking and understanding the language spoken so as the conversation with the calling party can continue as a natural conversation. A method has been presented which the aim is to fulfil the above stated aim.

The Prolog language which is a beanch of the first order logic language can be used as a combined formalism to signify the data which has been input into the system, output decision support rules, the reasoning system, background knowledge. The organized form of data is the reason behind it with significant dependency between the point that refined models are existing for knowledge in first order logic and the individual records (Shiue, 2009). The models can be called as the *inductive logic programming (ILP)*. Introduction of the first-order logic theories is the prime focus of the *inductive logic programming* from background knowledge and logic. Recenlty, there have been the establishment of the two branches of the ILP, named as the *nonmonotonic setting* and *normal setting,* the former one relates to the descriptive nature of the theories while the latter is about the predictive nature of the theories.

6.6. KNOWLEDGE PROCESSING IN CONTROL SYSTEMS

In atmposphere which are not completely organized, a number of big real time applications are needed. The methods of the problem solving are enavitable in the atmosphere of the uncertainity and deficiency of information. One of these applications is elevator group control. There are a number of probable scenarios consisting the state of elevators, accomplishment of earlier planned calls, present calls in the structure, and linking criteria of efficiency with new hall calls. There are a number of probable analogous timetables and when the new calls appear, there should be frequent revision of the already planned calls (Alsrehin et al., 2019). There is a requirement of the automated schedule creation procedure as it is not feasible to input all the schedules into the computer. For such scenarios, factory scheduling application can be utilized.

For real-time handling of knowledge, a matrix process was introduced by Looney, (2007) taking into account just the rules of the production with only one antecedent. In case when a number of antecedents are existing in a rule, his process does not protect the matrix. Moreover, it is hard to examine and anticipate the efficiency of the procedure to fulfil the deadlines because of the adopted system of chaining in case of multiple antecedents. In real-time scenarios this is a cruicial problem (Choudhary et al., 2009).

A substitute scheme on the basis of the kind of representation for the rules of the production are EUREK and RETE procedures. However, there is drawback of lack prediction feature which is necessary for the application in real-time.

Recently, a technique was developed that combines architectural primitives and methodology of problem solving to decrease the variation both the levels of problem-solving level and methodology level. They have displayed that by the use of this technique, the real-time and problem-solving can cohabit within a freely analyzable structure (Han et al., 1997).

The agenda is to provide a procedure of real-time knowledge processing (RTKP) on the basis of the disjunctive and conjunctive operators and matrices. In ths given procedure, the installation of a mechanism of the focus of attention is provided and its response time is assured. Following are some significant feature of the RTKP.

6.6.1. Preliminaries and General Definitions

A usual real-time knowledge processing which acts as direct digital control system is displayed in the Fig. 6.5. The module of the real-time knowledge processing is linked to the information receivers and sources. The receivers could be human users, actuators or computer programs In large connected systems, the source can be sensor linked to a human users, process or to a computer program. The primary concept behind real-time knowledge processing is that it collects information from the system, then processing of this information takes place with the stored knowledge and finally the processed information is sent to the system. A real-time knowledge processing shown in Figure 6.6 is carrying out supervisory control tasks (Padmanabhan & Tuzhilin, 2002). To assure temporal separation between conventional real-time tasks and time knowledge processing, knowledge-processing task is condensed within a server.

A typical real-time knowledge processing is categorized into four major units. First is the preprocessor module which is in control for the transmission of input data into the internal representation. Also, this module performs the mathematical handing (by creating transmission of variable), and to process the knowledge dependant on the task (Wiesner et al., 2011).

The translation into output information from the the internal representation is performed by the postprocessor module as needed by the procedure.

There are knowledge base and inference engine between these two modules. Internal knowledge repository comes at the end, coded in a practical arrangement. The inference engine processes the information given by the knowledge base and preprocessor module to produce the needed results.

The RTKP is well-defined by (La Rocca, 2012):

- Information which has been converted by the postprocessor module and information's internal representations given by the preprocessor
- K|nowledge base internal representation,
- Inference engine process

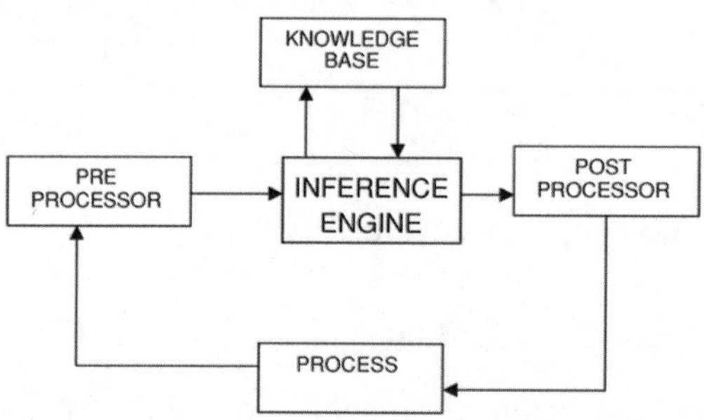

Figure 6.5. Typical RTKP structure in direct digital control

Source: https://link.springer.com/book/10.1007/978-3-540-34351-6

There are two major portions of the knowledge base: fact base and rule base. There is a truth value linked to a specific proposition in a fact which has been utilized to store the knowledge. There are set of terms in a fact base in which every term has a meaning linked to the procedure. Propositions are taken into account within the logic framework of preposition, however, there can also be fuzzy propositions (Jiang et al., 2019). The fact base is given in the suggested process is given by the fact vector in whuch every part is linked to the term and consists of its truth value.

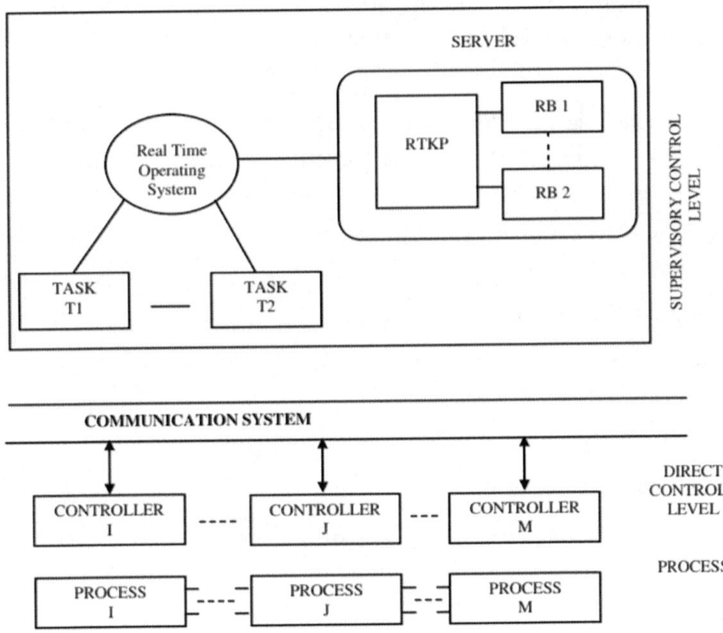

Figure 6.6. RTKP in supervisory control systems

Source: https://link.springer.com/book/10.1007/978-3-540-34351-6

There are two dissimilar represenations of the rule base. The initial representation is for the aim of the gaining of the knowledge and its analysis and it is named as the *virtual representation.* There are group of rules such as : antecedent then consequent, in which antecedent shows a disjunctive linkage of terms. There is another representation shich is about the coded type of the knowledge for the aim of processing (Hotho et al., 2001).

6.7. DATA MINING FOR MAINTENANCE OF COMPLEX SYSTEMS – A CASE STUDY

A huge collection of complicated data is produced by the maintenance and operation of novel systems embedded with the sensors including airplanes. Reduction in the cost of the operation may be achieved by the appropriate usage of the data to define and anticipate the failure of a part, which can further decrease the interruptions, and safetly of the overall system is enhanced. The domain of the knowledge discovery in databases has provided a number of methods from a huge collection of data (Zhang et al., 2017). Yet, there is not a single method which can cater the variation in data generated due to the complicated systems of maintenance and operation. A huge collection of data is used to take out valuable information which has been collected from 34 different airplanes in the period of three years. There have been various problems with the examination of the data which have been identified over some period of time (Letourneau et al., 1997). At first, varipus forms and sources of the data are taken into account. The data which is accessible includes: 1) failure and warning signals produced when specific scenarios happen, 2) a number of reports of sensor management which explain the condition of the airplane in various operations, 3) explanation of the issues in airplane including the maintenance activities performed for every problem. A number of sources are also available regarding background knowledge including empirical studies, training manuals and troubleshooting monitors. Quality and complications in the data also pose another problem. There are more than 100 parameters which need to be taken into account, and a number of paramters are anticipated to show a relationship of time-series, and there are also observations of inappropriate data forms, absent values out-of-range data. Atmospheric imapacts also observed to have effects on the sensor measurement which needs to be calibrated (Denecke & Nejdl, 2009).

To counter these problems, there are two possible research strategies: 1) grouping of the information from various sources given above to enhanve the quality of the data, 2) establishment of the approaches for the data preprocessing to cater the data quality and data complexity problems

The actions given below are addressed in the suggested data processing method: 1) data cleaning, 2) minimizing the influence of the environment on the data, and 3) classification of the occurrences so as machine learning methods can be utilized. Normalization methods independent of domains have been established in the research which utilize the examination of the variation (Letourneau.Matwin & Famili, 1998). The suggested method is a strong technique in decreasing the occurance of the false alarms generated due the accidental instabilities in the atmosphere as shown by the results from the experiment. Classification of the occurrences and data cleaning are the main aim of the research.

Another main feature of the research is concerns the various information sources which are obtainable in the maintenance and operation of the complicated systems. Concerning the following features, usage of the domain information has been applied: 1) study of the significance of the given models, 2) drawing out of the most suitable characteristics, and 3) accuracy enhancement. It is anticipated that the research would improve the procedure of knowledge discovery in the database, hence its application on the operation and maintence of the complicated sysmes would be easier (Raheja et al., 2006).

REFERENCE

1. Almana, A. M., Aksoy, M. S., & Alzahrani, R. (2014). A survey on data mining techniques in customer churn analysis for telecom industry. *International Journal of Engineering Research and Applications*, *4*(5), 165-171.
2. Alsrehin, N. O., Klaib, A. F., & Magableh, A. (2019). Intelligent transportation and control systems using data mining and machine learning techniques: A comprehensive study. *IEEE Access*, *7*, 49830-49857.
3. Amin, A., Anwar, S., Adnan, A., Nawaz, M., Alawfi, K., Hussain, A., & Huang, K. (2017). Customer churn prediction in the telecommunication sector using a rough set approach. *Neurocomputing*, *237*, 242-254.
4. Balasubramanian, M., & Selvarani, M. (2014). Churn prediction in mobile telecom system using data mining techniques. *International Journal of scientific and research publications*, *4*(4), 1-5.
5. Bandyopadhyay, D., & Sen, J. (2011). Internet of things: Applications and challenges in technology and standardization. Wireless personal communications, 58(1), 49-69.
6. Berkani, L. (2021). Decision support based on optimized data mining techniques: Application to mobile telecommunication companies. *Concurrency and Computation: Practice and Experience*, *33*(1), e5833.
7. Binti Oseman, K., Haris, N. A., & bin Abu Bakar, F. (2010). Data mining in churn analysis model for telecommunication industry. *Journal of Statistical Modeling and Analytics Vol*, *1*(19-27).
8. Bose, I., & Mahapatra, R. K. (2001). Business data mining—a machine learning perspective. *Information & management*, *39*(3), 211-225.
9. Camarinha-Matos, L. M., & Martinelli, F. J. (1998). Application of machine learning in water distribution networks. *Intelligent Data Analysis*, *2*(1-4), 311-332.
10. Choudhary, A. K., Harding, J. A., & Tiwari, M. K. (2009). Data mining in manufacturing: a review based on the kind of knowledge. *Journal of Intelligent Manufacturing*, *20*(5), 501-521.
11. Coussement, K., Lessmann, S., & Verstraeten, G. (2017). A comparative analysis of data preparation algorithms for customer churn prediction: A case study in the telecommunication industry. *Decision Support Systems*, *95*, 27-36.

12. Daskalaki, S., Kopanas, I., Goudara, M., & Avouris, N. (2003). Data mining for decision support on customer insolvency in telecommunications business. *European Journal of Operational Research*, *145*(2), 239-255.

13. Denecke, K., & Nejdl, W. (2009). How valuable is medical social media data? Content analysis of the medical web. *Information Sciences*, *179*(12), 1870-1880.

14. Dutt, A., Ismail, M. A., & Herawan, T. (2017). A systematic review on educational data mining. *Ieee Access*, *5*, 15991-16005.

15. Ewieda, M., Roushdy, M. I., & Shaaban, E. (2021). Review of Data Mining Techniques for Detecting Churners in the Telecommunication Industry. *Future Computing and Informatics Journal*, *6*(1), 1.

16. Eze, U. F., Onwuegbuchulam, C. J., Ugwuishiwu, C. H., & Diala, S. (2017). Application of data mining in telecommunication industry. *International Journal of Physical Sciences*, *12*(6), 74-88.

17. Farvaresh, H., & Sepehri, M. M. (2011). A data mining framework for detecting subscription fraud in telecommunication. *Engineering Applications of Artificial Intelligence*, *24*(1), 182-194.

18. Folasade, I. O. (2011). Computational intelligence in data mining and prospects in telecommunication industry. *Journal of Emerging Trends in Engineering and Applied Sciences*, *2*(4), 601-605.

19. Gheware, S. D., Kejkar, A. S., & Tondare, S. M. (2014). Data mining: Task, tools, techniques and applications. *International Journal of Advanced Research in Computer and Communication Engineering*, *3*(10) 3-19.

20. Gray, G. L., & Debreceny, R. S. (2014). A taxonomy to guide research on the application of data mining to fraud detection in financial statement audits. *International Journal of Accounting Information Systems*, *15*(4), 357-380.

21. Han, J., Koperski, K., & Stefanovic, N. (1997). GeoMiner: a system prototype for spatial data mining. *AcM sIGMoD Record*, *26*(2), 553-556.

22. Hashem, I. A. T., Chang, V., Anuar, N. B., Adewole, K., Yaqoob, I., Gani, A., ... & Chiroma, H. (2016). The role of big data in smart city. *International Journal of information management*, *36*(5), 748-758.

23. Hashmi, N., Butt, N. A., & Iqbal, M. (2013). Customer churn prediction in telecommunication a decade review and classification. *International Journal of Computer Science Issues (IJCSI)*, *10*(5), 271.
24. Hotho, A., Maedche, A., Staab, S., & Studer, R. (2001). SEAL-II - The Soft Spot between Richly Structured and Unstructured Knowledge. *Journal of Universal Computer Science*, *7*(7), 566-590.
25. Hung, S. Y., Yen, D. C., & Wang, H. Y. (2006). Applying data mining to telecom churn management. *Expert Systems with Applications*, *31*(3), 515-524.
26. Jadhav, R. J., & Pawar, U. T. (2011). Churn prediction in telecommunication using data mining technology. *International Journal of Advanced Computer Science and Applications*, *2*(2), 4-14.
27. Jiang, S., Wu, Z., Zhang, B., & Cha, H. S. (2019). Combined MvdXML and semantic technologies for green construction code checking. *Applied Sciences*, *9*(7), 1463.
28. Joseph, M. V. (2013). Data mining and business intelligence applications in telecommunication Industry. *International Journal of Engineering and Advanced Technology*, *2*(3), 525-528.
29. Keramati, A., Jafari-Marandi, R., Aliannejadi, M., Ahmadian, I., Mozaffari, M., & Abbasi, U. (2014). Improved churn prediction in telecommunication industry using data mining techniques. *Applied Soft Computing*, *24*, 994-1012.
30. Kirkland, J. D., Senator, T. E., Hayden, J. J., Dybala, T., Goldberg, H. G., & Shyr, P. (1999). The nasd regulation advanced-detection system (ads). AI Magazine, 20(1), 55-55.
31. Ko, S., Cho, I., Afzal, S., Yau, C., Chae, J., Malik, A., ... & Ebert, D. S. (2016). A survey on visual analysis approaches for financial data, 35(3), 599-617.
32. Kraljević, G., & Gotovac, S. (2010). Modeling data mining applications for prediction of prepaid churn in telecommunication services. *Automatika*, *51*(3), 275-283.
33. La Rocca, G. (2012). Knowledge based engineering: Between AI and CAD. Review of a language based technology to support engineering design. *Advanced engineering informatics*, *26*(2), 159-179.
34. Leite, R. A., Gschwandtner, T., Miksch, S., Kriglstein, S., Pohl, M., Gstrein, E., & Kuntner, J. (2017). Eva: Visual analytics to identify

fraudulent events. *IEEE transactions on visualization and computer graphics*, *24*(1), 330-339.

35. Li, Y., & Beaubouef, T. (2010). Data Mining: Concepts, Background and Methods of Integrating Uncertainty in Data Mining. *CCSC: SC Student E-Journal*, *3*, 2-7.
36. Liang, Y. H. (2010). Integration of data mining technologies to analyse customer value for the automotive maintenance industry. *Expert systems with Applications*, *37*(12), 7489-7496.
37. Liu, Y., Song, Y., Sun, J., Sun, C., Liu, C., & Chen, X. (2020). Understanding the relationship between food experiential quality and customer dining satisfaction: A perspective on negative bias. *International Journal of Hospitality Management*, *87*, 102381.
38. Looney, C. G. (2007). Fuzzy Data Mining in Higher Dimensions for Data Analysis. In *2007 IEEE International Conference on Information Reuse and Integration*, 4(1), 44-549.
39. Mahajan, V., Misra, R., & Mahajan, R. (2015). Review of data mining techniques for churn prediction in telecom. *Journal of Information and Organizational Sciences*, *39*(2), 183-197.
40. Manunza, L., Marseglia, S., & Romano, S. P. (2017). Kerberos: A real-time fraud detection system for IMS-enabled VoIP networks. *Journal of Network and Computer Applications*, *80*, 22-34.
41. Nadaf, M., & Kadam, V. (2013). Data mining in telecommunication. *Int. J. Adv. Comput. Theory Eng*, *2*, 92-96.
42. Nath, S. V., & Behara, R. S. (2003). Customer churn analysis in the wireless industry: A data mining approach. In *Proceedings-annual meeting of the decision sciences institute*, 561, 505-510.
43. Ngai, E. W., Hu, Y., Wong, Y. H., Chen, Y., & Sun, X. (2011). The application of data mining techniques in financial fraud detection: A classification framework and an academic review of literature. *Decision support systems*, *50*(3), 559-569.
44. Padmanabhan, B., & Tuzhilin, A. (2002). Knowledge refinement based on the discovery of unexpected patterns in data mining. *Decision Support Systems*, *33*(3), 309-321.
45. Qing, X., Li, W., Wang, Y., & Sun, H. (2019). Piezoelectric transducer-based structural health monitoring for aircraft applications. Sensors, 19(3), 545.

46. Raheja, D., Llinas, J., Nagi, R., & Romanowski, C. (2006). Data fusion/data mining-based architecture for condition-based maintenance. *International Journal of Production Research*, *44*(14), 2869-2887.

47. Ramageri, B. M., & Desai, B. L. (2013). Role of data mining in retail sector. *International Journal on Computer Science and Engineering*, *5*(1), 47.

48. Shaaban, E., Helmy, Y., Khedr, A., & Nasr, M. (2012). A proposed churn prediction model. *International Journal of Engineering Research and Applications*, *2*(4), 693-697.

49. Shadroo, S., & Rahmani, A. M. (2018). Systematic survey of big data and data mining in internet of things. *Computer Networks*, *139*, 19-47.

50. Shamsudin, M. F., Ali, A. M., Ali, A. M., & Shabi, K. S. (2019). Exploratory study of students'decision for enrolment at universiti kuala lumpur business school campus. *Humanities & Social Sciences Reviews*, *7*(2), 526-530.

51. Shiue, Y. R. (2009). Data-mining-based dynamic dispatching rule selection mechanism for shop floor control systems using a support vector machine approach. *International Journal of Production Research*, *47*(13), 3669-3690.

52. Sumathi, S., & Sivanandam, S. N. (2006). Data Mining in Telecommunications and Control. *Introduction to Data Mining and its Applications*,3(1), 615-627.

53. Toor, A. A., Usman, M., Younas, F., & Fong, A. (2020). A Robust Systematic Approach for Ensuring Optimal Telecom Service Delivery. *IEEE Communications Magazine*, *58*(8), 49-53.

54. Turhan, B., Kocak, G., & Bener, A. (2009). Data mining source code for locating software bugs: A case study in telecommunication industry. *Expert Systems with Applications*, *36*(6), 9986-9990.

55. Umayaparvathi, V., & Iyakutti, K. (2012). Applications of data mining techniques in telecom churn prediction. *International Journal of Computer Applications*, *42*(20), 5-9.

56. Wei, C. P., & Chiu, I. T. (2002). Turning telecommunications call details to churn prediction: a data mining approach. *Expert systems with applications*, *23*(2), 103-112.

57. Wiesner, A., Morbach, J., & Marquardt, W. (2011). Information integration in chemical process engineering based on semantic technologies. *Computers & Chemical Engineering*, *35*(4), 692-708.

58. Yang, Y., Xu, D. L., Yang, J. B., & Chen, Y. W. (2018). An evidential reasoning-based decision support system for handling customer complaints in mobile telecommunications. *Knowledge-Based Systems*, *162*, 202-210.

59. Zhang, Y., Ren, S., Liu, Y., & Si, S. (2017). A big data analytics architecture for cleaner manufacturing and maintenance processes of complex products. Journal of cleaner production, 142, 626-641.

Chapter 7

Data Mining In Security Systems

CONTENTS

7.1. INTRODUCTION

There are several advantages of data mining that can be employed to advance efficiency, quality of data, sales, and marketing. Moreover, it can also be employed in the security problems. We have discussed how the data mining tools could be employed to notice irregular behavior and interruptions in the system. In detecting fake behavior, data mining also has several applications (Thuraisingham et al., 2008). Humans can take advantage of these data mining applications, as they can pose a great threat to the privacy and security of individuals. There is also a risky side to mining. In this chapter, a summary of data mining in security and a real-time data mining-based interruption discovery system case study is shown (Cao, 2012).

Since progressively more sensitive data is being manipulated and stored online, the security of network systems is becoming increasingly important. To assist guard these systems, IDSs (Intrusion detection systems) have therefore become a serious technology.

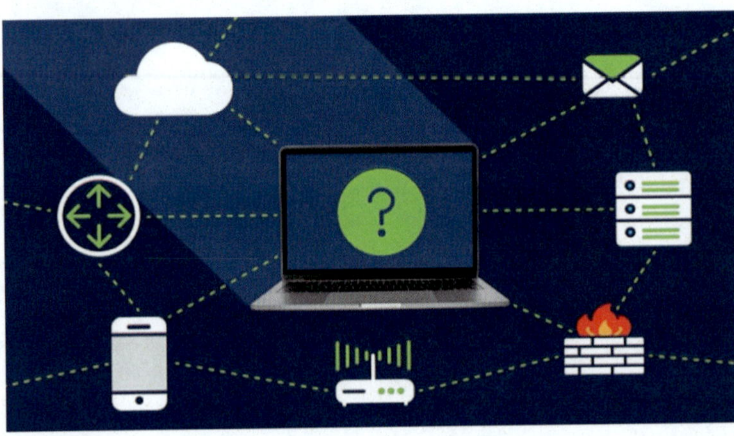

Figure 7.1. Components of network security

Source: https://www.cisco.com/c/en/us/products/security/what-is-network-security.html

Several intrusion detection systems are dependent on handmade signatures that are advanced through the manual encoding of proficient knowledge. In order to identify the signatures of attacks, these systems compare the activity on the system being observed to be identified (Gaber et al., 2019).

The main flaw in this method is that IDSs do not generalize to notice new attacks or attacks with unknown signatures. There has recently been a surge in interest in data mining-based methods to developing IDS detection models. These models simplify the known attacks and normal behavior in order to detect unexpected attacks. They can also be created more quickly and automatically than physically encoded models, which require time-consuming audit data processing.

For detecting intrusions, numerous successful data mining methods have been advanced, several of which perform and or improved than domain experts-engineered systems (Buczak & Guven, 2015).

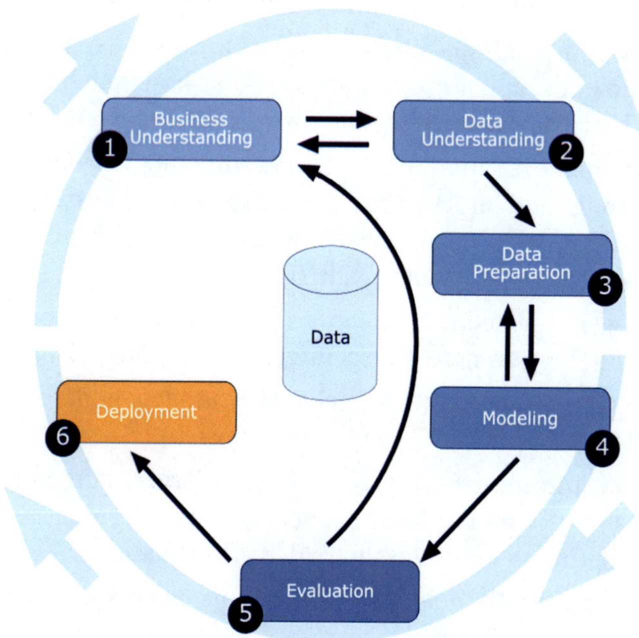

Figure 7.2. Association between Data mining and network security

Source:https://www.researchgate.net/publication/295907254_The_Role_of_Data_Mining_in_Information_Security

This chapter highlights several issues that arise while designing and deploying real-time data mining-based IDS and a summary of the research that tackles these issues.

7.2. ROLES OF DATA MINING IN SECURITY SYSTEMS

- Data mining is a useful tool for monitoring huge computer networks and ensuring their security. The data mining system accomplishes this by creating an outline of each network user's everyday actions (Molloy et al., 2010).
- Data mining tools are used to extract attributes and evaluate audit data that can separate regular operations from intrusions; fake anomalies are combined with normal and intrusion data to develop more operative misuse and irregularity, detection models.
- The computational charges of structures are examined to increase efficiency, and a multiple-model cost-based technique is utilized to build detection models with high accuracy and low cost.
- Adaptive learning methods are utilized to make incremental updates and model creation easier and to lessen the need for labeled data, unsupervised anomaly detection algorithms are employed (Vaidya et al., 2010).
- Intrusion detection systems (IDSs) have thus become a vital technology as network security becomes increasingly important.
- To detect unexpected attacks, IDS models generalize from both known assaults and normal behavior.
- Data mining-based IDSs have a larger percentage of false positives than outdated handcrafted signature-based approaches, rendering them useless in real-world situations.
- The anomaly detection algorithms investigate the usage of information-theoretic measures such as relative entropy, conditional entropy, entropy, information cost, and information gain to seizure intrinsic features of normal data and guide the development and evaluation of anomaly detection models (Vaidya et al., 2008).

7.3. DATA MINING AND SECURITY SYSTEMS

Vladimir Leonidovich Levin, an expert of the computer from St. Petersburg, Russia, broke into the Citibank electronic payments transfer network in June 1994. He transferred $10 million to accounts in Germany, Israel, California, Switzerland, the Netherlands, and Finland over 5 months. Although he was captured and the majority of the money was recovered, the episode highlighted the weakness of massive databases to computer hackers (Solove, 2008).

In today's corporate environment, incidents like the one defined above make the security of a company's computer system a critical concern. These computer networks, which can have thousands of computers and gigabytes of storage capacity, are monitored by system administrators and security officers.

Their job is daunting, especially since a security violation on one workstation could become a multimillion-dollar incident.

According to the Computer Emergency Response Team, a group of computer security professionals, barely 5% of firms whose security has been breached are even aware that they have been hacked. Although the basic information needed to detect an intrusion is frequently available in the audit data gathered by each machine, system administrators and security officers have simply too much of it to check each day. Even if they attempted, the great majority of the audit record would consist of entirely routine and insignificant activities (Fathima & Kiran, 2018).

Data mining provides a simple technique to keep track of these massive computer networks. A data mining system could flag worrisome occurrences for further investigation by system administrators by detecting abnormal actions in computer logs, allowing them to skip checking all of the routine everyday activities. The data mining system accomplishes this by creating a profile of each network user's everyday actions. Deviations from the intended design may be detrimental or rude and so would be detected.

Users learning novel programs would require the system to be flexible enough to accommodate for regular departures from expected behavior. According to a Purdue University study, a data mining system was able to identify a profiled user 99 percent of the time and differentiate between a profiled user and another user with around 94 percent accuracy (Kandogan & Haber, 2005; Adam et al., 2006).

7.4. REAL-TIME DATA MINING-BASED INTRUSION DETECTION SYSTEMS

We offer an overview of real-time data mining-based intrusion detection systems (IDSs) in this section, based on study of Lee et al. (2001) . Raleigh, North Carolina State University We concentrate on the challenges of deploying a data mining-based IDS in a real-time setting. We go over the methods for dealing with the three categories of problems: correctness, efficiency, and usability. Data mining tools are used to evaluate audit data and extract attributes that can separate regular operations from intrusions; fake anomalies are combined with intrusion or normal data to develop more operative anomaly and misuse detection, models (Lee et al., 2001).

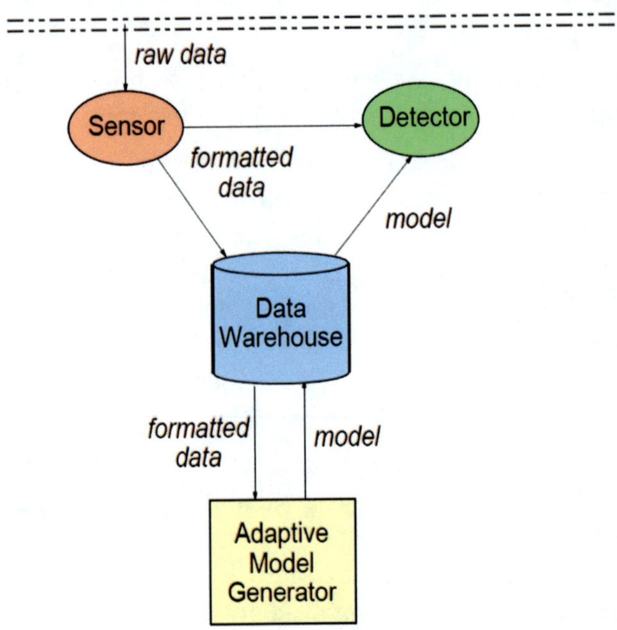

Figure 7.3. The Architecture of Data Mining-based intrusion detection system (IDS)

Source: https://www.researchgate.net/figure/The-Architecture-of-Data-Mining-based-IDS_fig1_2855820

The computational charges of features are examined to increase efficiency, and a multiple-model cost-based technique is utilized to build detection models with high accuracy and low cost. A distributed architecture for assessing cost-sensitive models in real-time is also presented. Adaptive learning methods are utilized to make model creation and incremental updates easier, and unsupervised anomaly detection algorithms are employed to lessen the need for labeled data. We also show a detector and sensor architecture, as well as data storage and model generating components (Raut

& Gawali, 2012). Successful data mining techniques, on the other hand, are insufficient to construct deployable IDSs. In spite of data mining-based IDSs' promise of improved detection generalization and performance ability, there are several important problems in their deployment and design. These issues can be classified into three categories: accuracy (detection performance), efficiency, and usability. Data mining-based IDSs (particularly anomaly detection systems) typically have larger false-positive rates than traditional handcrafted signature-based approaches, rendering them useless in real-world situations (Tian et al., 2003).

Furthermore, during both evaluation and training, these systems are inefficient. They are unable to analyze audit data and perceive intrusions in real-time as a result of this. To finish, unlike typical systems, these systems require a considerable quantity of training data and are substantially more sophisticated. These challenges must be addressed in order to deploy real-time data mining-based IDSs.

We explore many issues that arise when deploying and creating a real-time data mining-based IDS in this part and an overview of the research that tackles these issues. These issues are unrelated to the IDS's definite learning models or algorithms, and they must be solved to employ data mining techniques in a deployable system (Eskin et al., 2001). Each of these three types of difficulties must be addressed by a data mining-based IDS. Although there are compromises among these groupings, each can usually be dealt with independently. We provide the most important design aspects and categorize them according to the general concerns they address.

7.4.1. Accuracy

Determining how uncovering effectiveness or accuracy of these systems is monitored is serious to designing and implementing effective data mining-based IDS. Due to the differences in nature among a regular IDS and data mining-based system, assessment metrics for data mining-based systems must consider elements that are not significant for outdated IDSs.

Accuracy, at its most basic level, refers to how good an IDS notices attacks. An accuracy measurement has numerous important components. The detection rate, which is the %age of attacks that a system notices, is an essential factor (Dewa & Maglaras, 2016). The false-positive rate, which is the fraction of normal data that the system incorrectly considers intrusive, is another factor. In order to simulate a real deployment, these characteristics are often quantified by testing the system on a collection of data that were not seen during the training of the system. The detection rate and the false positive rate are inextricably linked. The detection rate against a false positive rate on a curve under different parameter values to create a ROC curve is one way to express this tradeoff. Examining the ROC curves of two IDSs is one way to assess their accuracy.

Only the little area of a ROC curve matching suitable low false positives is of interest, as only a low false-positive rate can be tolerated in a deployable system. Handcrafted approaches usually feature a defined detection brink and function at a consistent detection rate across a range of false-positive rates (Panda & Patra, 2008). We can suppose that the ROC curve is a straight line at each detection level in a ROC curve. Data mining-based systems have the advantage of being able to detect novel assaults that traditional methods often overlook.IDSs based on data mining are only beneficial if they have a greater detection rate than a handmade technique with a low false-positive rate. The goal is to construct a data mining-based IDS that can outperform handcrafted signature-based systems at the allowed false positive rate using this framework. To increase the performance of data mining-based IDSs, we developed and used several algorithm-independent approaches (Sabri et al., 2011). This section focuses on a few specific approaches that have been empirically demonstrated to be effective. We begin by presenting a basic method for identifying elements from audit data that aid in distinguishing attacks from regular data. Any detection model-building technique can then use these attributes. Then, in order to reduce the false-positive rate of anomaly detection systems, we offer a method for creating artificial anomalies. We can improve the accuracy of these ID models by creating false abnormalities, according to the research.

To finish, we show how to combine misuse and anomaly detection models in one technique. Models for detecting misuse and anomalies are often trained and deployed in perfect isolation from one another. We can enhance the overall detection rate of the system by merging the two types of models, according to the research, without sacrificing the benefits of either detection approach (Singh et al., 2011).

7.4.2. Feature Extraction for IDS

Two fundamental axioms of intrusion detection are that system actions may be observed, such as through auditing, and that different indications can be used to discriminate between intrusive and normal activity. The evidence retrieved from raw review data is referred to as features, and these features are employed to create and test intrusion detection models. Deciding what indication can be extracted from raw audit data that is most valuable for analysis is known as feature extraction. As a result, feature extraction is a crucial phase in the development of an IDS. Strong detection performance requires a set of features whose values in regular audit data deviate significantly from the values in intrusion records (Karimi et al., 2016).

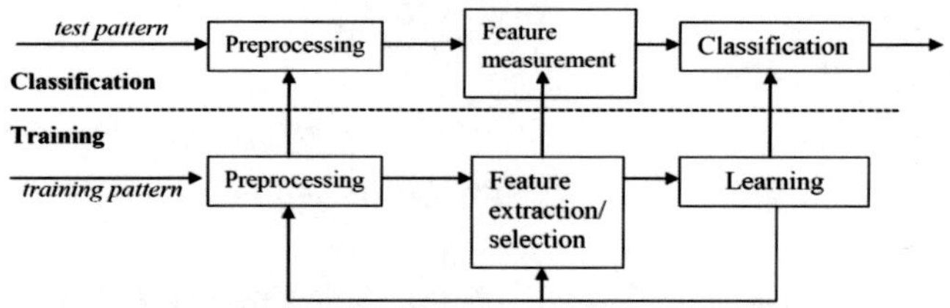

Figure 7.4. Model of a statistical pattern recognition

Source: https://www.semanticscholar.org/paper/Feature-Extraction-Methods-for-Intrusion-Detection-Nguyen-Franke/f9674966b22a2969d0807ae566442701cc8ae23a/figure/0

For picking and generating features from audit data, we built a set of data mining techniques. The raw (binary) audit data is first processed and brief into distinct records, including numerous fundamental properties, such as source, durations, timestamps, and destination ports and IP addresses, and error condition flags in the situation of network traffic. The common patterns indicating correlations among the attributes and often co-occurring events across numerous records are then computed using specialized data mining techniques (Yan & Han, 2018). A pattern is commonly defined as A, B, C, D, which means that occurrences A and B are followed by events C and D with a high degree of inevitability and a high frequency in the data. The "consistent" patterns are associated with routine activities, whereas the "unique" patterns connected with an interruption are recognized and evaluated to build extra features for connection records. It has been demonstrated that created features can effectively distinguish intrusion from typical behavior. The built features are more empirically based and consequently more objective than expert knowledge when using this approach. According to the 1998 DARPA Intrusion Detection Evaluation results, an IDS model built utilizing these techniques was one of the top-performing of all the systems tested (Kasongo & Sun, 2020).

Let us use the SYN flood attack as an example. When conducting this attack, an attacker uses a large number of faked source addresses to initiate a large number of connections to a target host's port that never gets fully established. By first encoding the patterns into numbers and then computing "difference" scores, we compared the patterns from the 1998 DARPA data set that contain SYN flood attacks with the patterns from a "baseline" regular data set. The highest "intrusion-only" score goes to the following pattern, which is a frequent episode (Schadt et al., 2001):

"*(flag = S0, service = http, dst host = victim), (flag = S0, service = http, dst host = victim) → (flag = S0, service = http, dst host = victim)* [0.93, 0.03, 2]."

This indicates that 93 percent of the time after two HTTP influences with the S0 flag are complete to host prey, the 3rd parallel connection is made within 2 seconds of the first of these two; this pattern happens in 3% of the data. As a result, the pattern characteristics are parsed by the feature creation algorithm: "a count of connections to the same dst host in the previous 2 seconds," and "the percentage of those that have the same service, and the % of those that have the S0 flag" among these connections (Asadi & Lin, 2013). The even connection records contain values close to 0 for the 2 "percentage" attributes, but the syn flood connection records had values overhead 80%. This means that after two HTTP connections with the S0 flag are made to the host victim, the third similar connection is made within two seconds of the first of these two, and this pattern occurs in 3 percent of the data. The feature construction method parses the pattern characteristics as follows: "a count of connections to the same dst host in the preceding 2 seconds," "the proportion of those that have the same service, and the percentage of those that have the S0 flag" among these connections. The two "percentage" properties in ordinary connection records are close to 0, whereas the syn flood connection records have values above 80% (Hao et al., 2019).

7.4.3. Artificial Anomaly Generation

Making the learner identify boundaries among unknown and known classes is a major challenge when utilizing machine learning approaches for anomaly detection. A machine-learning system will only identify boundaries that distinct various known classes in training data because there are no examples of anomalies in the data (Fan et al., 2004).

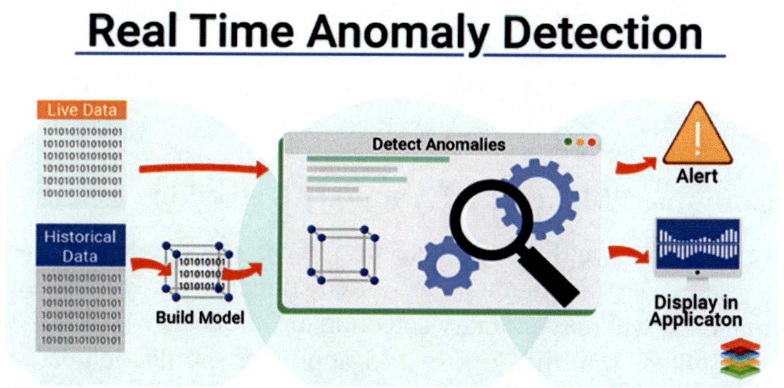

Figure 7.5. Real-time anomaly detection phenomenon

Source: https://www.xenonstack.com/blog/real-time-anomaly-detection

A machine-learning algorithm does not draw a line between known and unknown data. We show how to use artificial anomaly creation to teach traditional learners how to spot anomalies. Artificial anomalies are introduced into the training data to assist the learner in identifying a border around the actual data (Stolfo et al., 2001). The class label anomaly is applied to all manufactured abnormalities. The method for creating artificial anomalies focuses on "near misses," or instances that are similar to known data but not in the training data. Near misses can be safely presumed to be abnormal because we assume the training data is representative.

We assume that the decision border between anomalous and known cases is near to the available data because we do not know where it is. A helpful heuristic for generating fake anomalies near known data is to modify the value of one feature of an example to a value that does not happen in the data whereas leaving the other features unchanged (Gómez et al., 2003).

Instance space may have some poorly filled regions of known data. The thin regions can be compared to little islands in the ocean, while the dense portions can be compared to enormous islands. Learning algorithms are typically geared toward discovering more generic models to avoid overfitting. We do not want models to be very generic when predicting these known classes because we only have known data. That is, we don't want to end up with a condition where thin regions are clustered together to form dense regions, resulting in single vast regions covered by excessively general models (Hwang et al., 2007). Artificial anomalies can be created at the edges of these thin areas to force the learning algorithm to find the distinct boundaries that separate the regions from the rest of the occurrence space. To put it another way, we want to collect data that will increase these scarce areas.

7.4.4. Combined Misuse and Anomaly Detection

Anomaly detection and abuse detection have always been treated as different issues. Misuse methods are often trained over labeled intrusion and normal data, whereas anomaly detection techniques are typically trained over normal data. Intuitively, a hybrid approach should outperform two separate models; additionally, it offers apparent efficiency advantages in both model training and deployment (Zarrabi & Zarrabi, 2012). We apply the fake anomaly creation method to construct a single model that is both a misuse and anomaly detection approach. This enables us to employ typical supervised inductive learning approaches to detect both anomalies and misuse simultaneously (Kim et al., 2014).

We train a single model from data that includes both records consistent with intrusions and normal records. We also use the DBA2 algorithm to create false anomalies and train the system on the mutual data set. Anomalies and intrusions can both be detected using the learned model.

For anomaly detection and combined abuse, we learn a single ruleset. A connection can be classified as normal, known as intrusion classifications, or anomaly using the ruleset. We aggregate invasions into several small clusters in order to test this combination method. By iteratively adding each cluster to the data set and regenerating artificial anomalies, we construct several data sets (Agrawal & Agrawal, 2015).

This is done to mimic the real-world process of creating and determining new incursions and adding them into the training set. We create models that include abuse rules for known intrusions in the training data, anomaly detection rules for unidentified intrusions in left-out clusters, and normal behavior rules.

7.4.5. Efficiency

Since the learning procedures must handle massive volumes of file-away audit data, detection models are typically created offline in distinctive data mining systems for intrusion detection. Offline intrusion detection is a logical application for these models (Aydın et al., 2009). We'll talk about making data mining-based ID models perform well for real-time intrusion detection in this part.

Contrary to offline IDSs, real-time IDSs have the goal of detecting intrusions as soon as possible. As a result, the detection model's competence is a critical factor to consider. Because data mining-based models are built employing offline data, they indirectly accept that all relevant activities have been completed when an event is reviewed, and all features have meaningful values accessible for model testing (Pan et al., 2105; Smys et al., 2020). As a result, in real-time, if we apply these models without making any changes, an event will not be inspected until all relevant information has been reached and been brief and all statistical and temporal aspects have been computed. Under real-time restrictions, this scheme may flop badly. When an event stream's volume is high, it takes a long time to process the event recordings from the previous n seconds and calculate numerical features. Several later events may have ended when the "current" event is eventually examined through the model. In other words, intrusion detection takes a long time. Inappropriately, intruders frequently utilize DoS assaults, which create a significant quantity of traffic in a short period, to first overload an IDS and then use the detection delay to quickly carry out their nefarious objective (Hajisalem & Babaie 2018). They can, for example, acquire control of the host on which the IDS is installed, effectively nullifying the usefulness of intrusion detection.

To speed up model evaluation, it is required to investigate the time delay associated with computing each feature. A feature's time delay includes the time spent computing it and the time spent waiting for it to be ready. In-network auditing, for example, the total time of a network connection can only be determined after the last packet of the connection has arrived, although the destination host of a connection can be determined by inspecting the header of the first packet (Yan et al., 2009).

The computational cost of an intrusion detection model, which is the quantity of the time delays of the characteristics utilized in the model, is its efficiency in terms of cost analysis. We may categorize features utilized for network intrusion detection into four cost levels based on the feature construction methodologies outlined in the preceding section (Rajendran et al., 2015).

- This packet can be used to compute level 1 features, such as the service.

- At any time, level 2 features can be computed during the connection state's lifetime.
- At the end of a connection, level 3 features can be computed by employing only the data about the connection being investigated, such as the total number of bytes sent from source to last stop.
- Four features can be computed at the end of a connection level, but they need data from many previous connections. These are the most time-consuming and statistical features to compute.

To make estimating the cost of a rule easier, we give level 1 features a cost of 1, and level 2 features a cost of 5, level 3 features a cost of 10, and level 4 features a cost of 100. These cost estimates are extremely close to the actual measurements obtained after intensive real-time testing (Mohan et al., 2019). Nevertheless, we discovered that the cost of calculation level 4 characteristics is proportional to the number of connections monitored by the IDS inside the computation time window because they need repetition of the entire collection of new connections (Ghosh et al., 2014).

7.4.6. Cost-Sensitive Modeling

To lower the computational cost of IDS, detection rules must employ low-cost characters as frequently as probable while keeping the necessary level of accuracy. We suggest a multiple rule set strategy, in which every rule set takes advantage of various cost level features. IDS always evaluate low-cost rules first, and high-cost rules are applied only when low-cost rules cannot be forecast with enough accuracy (Lee et al., 2002).

As described in the earlier method, we employ four different degrees of costs to calculate features in the domain of network intrusion detection. Costs 1, 5, and 10 have unique features, whereas cost 100 has a single lookup of all the connections in the last n seconds. We utilize the following multiple rules set technique with the overhead expenses and aim in mind (Sahin et al., 2013).

- We start by creating numerous training sets T_1–4 with various feature subsets. Only cost one feature is used in T_1. T_2 employs the features of costs 1 and 5, and so on until T_4 employs all of the available characteristics.
- Sets of rules R_1–4 are acquired through the use of their specific training sets. R_4 is taught as an ordered rule set. The rules are checked in order. The rule for the most common class, i.e.,

normal, is usually placed first. Because it may contain the most expensive features, it is efficient. R_1−3 are cultured as unordered rule sets because they will contain accurate rules for categorizing normal connections.

- A precise extent is required. [Precision is a term that refers to how accurate a prediction is. Precision is computed for every ruler, except for rules in R_4, as p = |PW||P|, where P is the set of predictions with label I and W is the set of all instances with label I in the data set (Strasburg et al., 2009).
- For each class, a threshold value τ_I is calculated, which sets the acceptable accuracy required for any rule set except R_4 to make a classification.

In a real-time implementation, the feature calculation and rule assessment proceed as follows:

- For the connection under investigation, all cost one feature used in R_1 is computed. Following that, R_1 is analyzed, and a prediction $time_I$ is made.
- The prediction i is fired if pr >τ. No more features are calculated in this situation, and the system moves on to the next linking. Else, R_2's additional features are computed, and R_2 is assessed in the same way that R1 is (Telikani & Gandomi, 2019).
- With R_3 assessment stays, tracked by R_4 till the prediction is made.
- When R_4 is reached, features are calculated as required, and the ruleset is evaluated from top to bottom. R_4 does not need any firing situations to be evaluated, and it will always produce an estimate.

They employed data from a 1998 DARPA evaluation in the studies. Lee et al. (2002) published the entire experimental setup and findings in 2000. To summarize, the multiple model technique can lower computing costs by up to 97 percent without sacrificing predictive accuracy, where the cost of checking a link in the entire computational cost of all exceptional features is employed before making a forecast. If more than one feature with a cost of 100 is employed, the cost is only counted once because they can all be calculated in one cycle over the database of new connections (Kirkpatrick, 1978; Conrad, 1988).

7.4.7. Distributed Feature Computation

We developed a system that can evaluate some cost-sensitive models in real-time. This system creates connection records using a sensor to collect lightweight or "basic," characteristics from raw network traffic data and then offloads model assessment and higher-level feature computation to a separate entity named JUDGE (Liang et al., 2019). The reason for outsourcing this assessment and computation is that it is quite expensive, and we do not want the sensor to be overburdened.

JUDGE makes use of models that have been erudite using the methods previously discussed. There is a series of models, each of which uses more expensive features than the one before it. As more rudimentary features become available, JUDGE evaluates models and computes higher-level features at various points in a connection (Jabez & Muthukumar, 2015).

When the connection's state changes, the sensor notifies JUDGE of new feature values or updates to feature values that have been retained during the connection's life. When an "exception" event occurs, sensors update some feature values. Exceptions are particular events that should cause the value of a feature to be updated instantly (Li et al., 2109). If two fragmented packets arrive and the defragmentation offsets are proper, the bad frag offset characteristic must be adjusted right away.

JUDGE figures the list of features existing for the specified connection with each state change and exception occurrence. Higher-level features are computed, and that model has evaluated if the set of features is a suitable subset of the set of light-weighted features used by one of the ID models. The rationale for deciding whether a prediction is made follows the same pattern as previously discussed (Iman & Ahmad, 2020).

Once JUDGE has made a prediction, the entire connection record, including the label, is introduced into a data warehouse, as stated in the initial system design. Although the protocol for communication among the JUDGE and sensor allows any sensor that extracts features from a data stream to be utilized, we have currently implemented this system utilizing NFR's Network Flight Recorder as the sensor.

7.4.7.1. Usability

An IDS based on data mining is far more complicated than a typical system. The fundamental reason for this is that data mining systems need a large amount of data to learn. Many active research topics have resulted from the need to reduce the complexity of data mining systems (Singhal & Jajodia, 2006).

For starters, managing both historical and training data sets is a complex undertaking, especially if the system deals with various data types. Second, models must be updated when new data has been analyzed. Update models by retraining over all available data are unrealistic since retraining might take weeks, if not months, and new models are required instantly to assure system security. To modify a model to new knowledge, some mechanism is required. Third, numerous data mining-based IDSs are challenging to implement because they require a large quantity of labeled training data that is clean (Patel et al., 2012).

Classically, in the data, the attacks must be manually labeled for signature detection models or eliminated for anomaly detection models to be trained. Cleaning training data by hand is time-consuming, especially in the case of large networks. We must lower the amount of clean data required by the data mining process to reduce the cost of establishing a system.

We propose a solution to each of these issues. We employ adaptive learning, which is a general mechanism for adding new data to a model without retraining it. We use unsupervised anomaly detection, a type of intrusion detection technique that does not rely on labeled data. We will show you a system architecture that automates model and data management in the next part (Patond & Deshmukh, 2014).

7.4.7.2. Adaptive Learning

We suggest using bands of sorting models as a broad, algorithm-independent strategy for adapting current models to discover newly developed patterns. The goal is to make both deployment and learning more efficient. In actuality, when a new type of intrusion is detected, being able to quickly adapt an existing detection system to identify the new assault, even if the alteration is temporary and may not catch the new attack, is very desirable (HU et al., 2007; Yu, 2010).

Simultaneously, once we have at least some form of defense, we can explore for possible better ways to notice the attack, which will require recomputing the uncovering model, which could take a long time. We may decide to change the momentary model better after a better model has been generated. We want a "plug-in" technique for such purposes, in which we proficiently construct a simple model that is solely good at detecting new intrusions and plug or attach it to current models to permit uncovering of new intrusions (Borkar et al., 2019).

- **If** (H_1 (x) = *normal*) V (H_1 (x) = *anomaly*) **then**
  - **if** H_2 (x) = *normal*
    then *output* ← H_1 (x) (*normal or anomaly*)
  - **else** *output* ← *new_intrusion*
- **else** *output* ← H_1**(x)**

Figure 7.6. Ensemble-based Adaptive Learning Configuration

Source: https://link.springer.com/book/10.1007/978-3-540-34351-6

We create a lightweight classifier for the novel pattern quickly and efficiently. The current principal detection model is unaffected. When the old model detects an anomaly, the data record is submitted to the new classifier for classification. Both the new and old classifiers work together to make the last prediction (Yu et al., 2005). Creating a monolithic model for all established patterns and anomalies takes substantially longer than computing the new classifier.

Given a new classifier, H_2, a present classifier H_1, is trained from data comprising standard records and records matching to the new incursion in one such configuration. The old IDS model is referred to as H_1, whereas a new model trained particularly for a new or lately found attack is referred to as H_2. To compute the outcome, the choice rules in Fig.25.1 are examined. The real model-building algorithm is unaffected by this strategy (Lui et al., 2005). A neural network, a rule-based learner, a decision tree, and so on can all be used as classifiers.

We tried several combinations to see how effective this strategy is. The suggested method's training cost (as determined by cost-sensitive modeling) is nearly 150 times lower than learning a monolithic classifier taught from all available data, and their accuracy is nearly equal (Lee et al., 2000).

7.4.7.3. Unsupervised Learning

In order to generate good detection models, traditional model-building techniques often require a huge amount of labeled data. One of the most difficult aspects of implementing a data mining-based IDS is classifying system audit data for usage through these algorithms. The data must be appropriately tagged as attack or normal for abuse detection systems. The data must be validated to guarantee that it is normal for the anomaly detection system, which involves the same effort (Ariafar & Kiani, 2017). The expense of labeling the data must be incurred for each deployment of the system because models (and data) are specific to the context in which the training data was collected.

From the collected data, we'd like to develop detection models deprived of having to label it manually. Since the data would not be essential to be labeled, the deployment cost would be considerably reduced. We'll need a new type of model-building algorithm to create these detecting models. These model-building techniques can develop a detection model from unlabeled data. These techniques are referred to as unsupervised irregularity detection algorithms.

In this part, we discuss the challenge of unsupervised difference identification and its relation to the statistical problem of outlier detection. We give a brief review of two unsupervised anomaly detection methods that have been used to detect intrusions (Jianliang et al., 2009).

Anomaly detection over noisy data is another name for these algorithms. We do not want to manually confirm that the audit data collected is totally clean; hence the algorithm must handle noise in the data.

Two main norms about the data drive unsupervised anomaly detection methods, both of which are appropriate for intrusion detection. The first hypothesis is that anomalies are extremely uncommon. This is because typical system usage far outstrips the occurrence of invasions. As a result, the attacks account for only a small percentage of the real data (Bhavsar & Waghmare, 2103). The second premise is that the anomalies differ from the normal elements quantitatively. This corresponds to the fact that attacks differ significantly from typical usage in intrusion detection.

Anomalies show up as outliers in the data set because they are extremely infrequent and statistically different from the regular data. As a result, the problem of identifying attacks can be recast as an outlier detection problem. In the discipline of statistics, outlier detection has gotten much attention (Ibrahim et al., 2013).

In intrusion detection, instinctively, if the ratios of assaults to usual data are small enough, the attacks will show up compared to the background of normal data since they are distinct. As a result, the assault can be detected inside the data set (Gu & Zhang, 2009).

We tested two different types of unsupervised anomaly detection methods, each for a different kind of data. We used a probabilistic-based unsupervised anomaly detection technique and a clustering-based unsupervised anomaly detection algorithm for network traffic to create detection models for system calls.

Estimating the likelihood of each element in the data is one of the probabilistic ways to detecting outliers. We divide the data into two categories: normal and anomalous items. We compute the most likely partition of the data using a probability-modeling approach applied to the data (Aziz et al., 2012).

By clustering the data, the clustering approach detects outliers. The assumption is that because there is so much of it, the standard data will group. Standard data and anomaly data do not group since they are so dissimilar. Because there is so slight inconsistent data compared to normal data, the anomalous data will end up in small clusters after clustering. After clustering the data, the program identifies the least groups as anomalies (Desale et al., 2015).

7.4.8. System Architecture

The entire system architecture is built to backing a data mining-based IDS that meets the requirements outlined in this section. The architecture comprises a data warehouse, detectors, sensors, and a model-producing component, as illustrated in Fig.25.2. Data gathering, analysis and sharing, and data preservation, and model building and distribution are all supported by this architecture.

The system is built to work regardless of sensor data type or model illustration. An arbitrary number of structures can be found in a single piece of sensor data. Every feature can be discrete or continuous, symbolic or numerical, and can be continuous or discrete (Idhammad et al., 2108). A model can be everything in this outline, from a neural system to a set of rules to a probabilistic model. An XML encoding is utilized to deal with this heterogeneity, allowing apiece component to simply interchange data and models.

The work of the Common Intrusion Detection Framework (CIDF, supported by DARPA) and the more recent Intrusion Detection Message Exchange Format in standardizing protocols and message formats for IDS communication and collaboration affected the design (IDMEF by the Intrusion Detection Working Group of IETF, the Internet Engineering Task Force) (Bridges & Vaughn, 2000). IDSs can securely transmit attack information, encoded in common formats, using CIDF or IDMEF to detect distributed attacks collectively.

In the architecture, model and data replaced among the components are

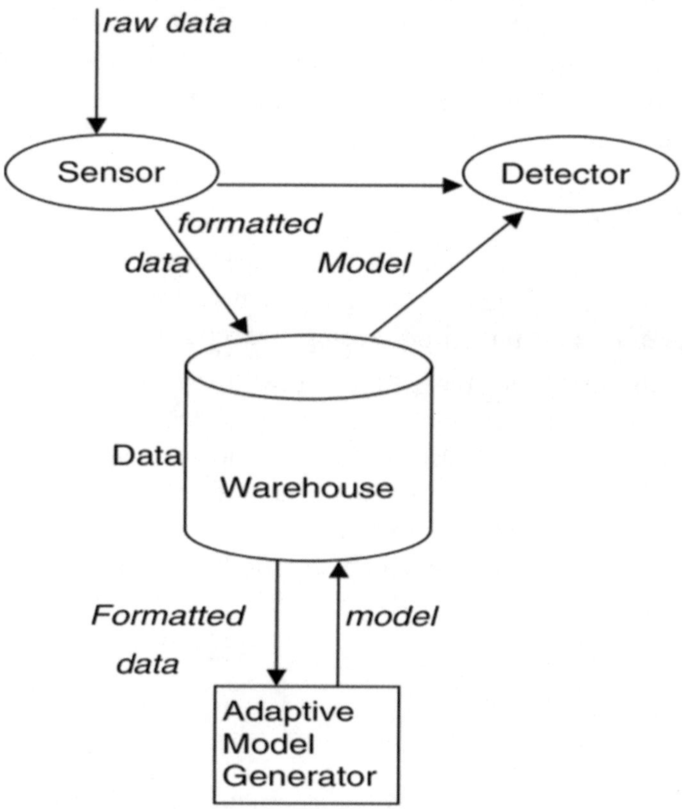

Figure 7.7. The Architecture of Data Mining-based IDS

Source: https://www.researchgate.net/figure/The-Architecture-of-Data-Mining-based-IDS_fig1_2855820

encoded in our normal message format, which can be irrelevantly mapped to either IDMEF or CIDF formats. The architecture's main benefit is its great scalability and performance. Such as, entirely components can be on a similar local network, in which situation the workload is split among them; or the components can be on separate networks, in which case they can collaborate with other IDSs on the Internet (Hajimirzaei & Navimipour, 2019). The components represented in Fig.25.2 are further described in the sections that follow.

7.4.8.1. Sensor

Sensors collect raw data from an observed system and calculate features that can be used in model evaluation. Sensors shield the remainder of the IDS from the low-level features of the monitored target system. This is accomplished by requiring all sensors to use the Basic Auditing Module (BAM) framework (Kshirsagar et al., 2012). Features are calculated from raw data and encoded in XML in a BAM.

7.4.8.2. Detectors

From sensors, detectors gather processed data and analyze it using a detection model to see if it's an assault. In addition, the detectors submit the results to the data warehouse for further processing and reporting.

Several detectors can be used to keep an eye on the same system. Workloads can be spread among different detectors to examine events simultaneously. There may also be a "back-end" detector that uses highly cultured models for trend or correlation analysis, as well as multiple "front-end" detectors that detect intrusions quickly and easily (Wu & Yen, 2009). The front-end detectors must keep up with high-volume traffic, high-speed and pass data to the back-end detector, which performs a more detailed and time-consuming examination.

7.4.8.3. Data warehouse

The data warehouse stores data and models in a concentrated location. With the availability of a database, multiple components can use the same piece of data asynchronously, like manually labeling off- and line training, which is one benefit of a centralized data repository. Data can be manipulated simultaneously by the same type of components, such as several sensors (Qi & Dong, 2012). The "stored process calls" feature of relational databases allows for the simple implementation of complex calculations, like effective data sampling on the server carried out automatically.

A single SQL question can also regain an unlimited amount of sensor data. The distribution of detection models can be pulled or pushed.

Data from various sensors can also be integrated using the data warehouse. The detection of intricate and large-scale assaults becomes achievable by matching data/results from different IDSs or data acquired over a longer period (Lee & Soh, 2003).

7.4.8.4. Model Generator

The model generator's main goal is to speed up developing and deploying new intrusion detection models. In this architecture, an attack detected as an irregularity may have its model data managed through the model generator, which automatically produces a model that can perceive the new distributes and intrusion it to the detectors using the archived normal and intrusion data sets from the data warehouse (Xiangrong et al., 2001). Unsupervised anomaly detection techniques are particularly valuable since they may work on unlabeled data gathered directly by sensors.

We successfully finished a data mining and CIDF-based IDS prototype implementation. A data mining engine with feature extraction and machine learning algorithms assists as the model generator for numerous detectors in this system. It collects audit data from a detector for rare events, calculates designs from the data, equivalences them to usual past patterns to find "exceptional" intrusion patterns, and builds features, therefore. The detection model is then computed using machine learning algorithms, programmed as GIDO, and delivered to all detectors (Raut & Gawali, 2012). The standard intrusion detection models focused on much of the design and implementation work (CISL). After obtaining audit data, the generator can develop and distribute new effective models in preliminary tests.

7.4.8.5. Related Work

The study covers various topics, including machine learning, data mining, and intrusion detection. In this part, we compare and contrast our techniques to similar endeavors.

DC-1 constructs a cellular phone fraud detector by primarily appealing a sequence of operations for creating features (indicators). Because there is no common record format for connection or session records, we are confronted with a more complex challenge. For individual records and various services and connections, we must develop statistical and temporal aspects (Pan et al., 2015). We are demonstrating many logical units that play various roles and whose activity is meticulously recorded. The task of extracting these from a massive and overwhelming stream of data enhances the problem's complexity.

At SRI, a method developed in the Emerald system is very similar to unverified model generation. Emerald builds normal detection models from history records and likens novel instance distributions to historical distributions. Disparities in the distributions indicate an incursion. One issue with this method is that intrusions in previous distributions may cause the system to miss comparable invasions in unobserved data (Ashraf et al., 2018).

Adaptive intrusion detection is related to automatic model generation. Pan et al. (2015) use inductively produced sequential patterns to conduct adaptive real-time anomaly detection. On adaptive intrusion detection, Sobirey's work is utilizing a proficient system to gather data from check sources is also relevant.

A variety of methods for developing anomaly detection models have been offered. Stephanie Forrest demonstrates how contiguous sequences and look-ahead pairs can be used to represent normal sequences. Bhangoo and Helman provide a statistical technique for identifying more common sequences in incursion data than in normal data. Lee et al. employ a decision tree to train a prediction model that is applied to normal data (Dhanabal & Shantharajah, 2015). To model normal data, Schwarzbard and Ghosh employ neural networks. Brodley and Lane look at user profiles and compare activity during an incursion to activity during normal use to seek anomalies in unlabeled data.

Due to the demand from application areas like cost-sensitive modeling and intrusion and fraud detection, medical diagnosis is a hot topic in machine learning and data mining. Several strategies for designing buildings that are optimized for specific cost criteria have been offered. We investigate the principles underpinning these broad procedures in our research and propose new approaches based on cost models particular to IDSs. The IETF Intrusion Detection Exchange Format initiative and the CIDF endeavor are both linked to intrusion data depiction (Hossain et al., 2003).

REFERENCES

1. Adam, N. R., Atluri, V., Koslowski, R., Grossman, R., Janeja, V. P., & Warner, J. (2006). Secure interoperation for effective data mining in border control and homeland security applications. In *Proceedings of the 2006 international conference on Digital government research*, 3(2), 124-125.
2. Agrawal, S., & Agrawal, J. (2015). Survey on anomaly detection using data mining techniques. *Procedia Computer Science*, *60*, 708-713.
3. Amini, M., Jalili, R., & Shahriari, H. R. (2006). RT-UNNID: A practical solution to real-time network-based intrusion detection using unsupervised neural networks. *computers & security*, *25*(6), 459-468.
4. Ariafar, E., & Kiani, R. (2017). Intrusion detection system using an optimized framework based on datamining techniques. In *2017 IEEE 4th International Conference on Knowledge-Based Engineering and Innovation, 13(1),* 785-791.
5. Asadi, N., & Lin, J. (2013). Document vector representations for feature extraction in multi-stage document ranking. *Information retrieval*, *16*(6), 747-768.
6. Ashraf, N., Ahmad, W., & Ashraf, R. (2018). A comparative study of data mining algorithms for high detection rate in intrusion detection system. *Annals of Emerging Technologies in Computing (AETiC), Print ISSN*, 3(1), 2516-0281.
7. Aydın, M. A., Zaim, A. H., & Ceylan, K. G. (2009). A hybrid intrusion detection system design for computer network security. *Computers & Electrical Engineering*, *35*(3), 517-526.
8. Aziz, A. S. A., Salama, M. A., ella Hassanien, A., & Hanafi, S. E. O. (2012). Artificial immune system inspired intrusion detection system using genetic algorithm. *Informatica*, *36*(4), 4-10.
9. Beghdad, R. (2008). Critical study of neural networks in detecting intrusions. *Computers & security*, *27*(5-6), 168-175.
10. Bhavsar, Y. B., & Waghmare, K. C. (2013). Intrusion detection system using data mining technique: Support vector machine. *International Journal of Emerging Technology and Advanced Engineering*, *3*(3), 581-586.
11. Borkar, G. M., Patil, L. H., Dalgade, D., & Hutke, A. (2019). A novel clustering approach and adaptive SVM classifier for intrusion detection

in WSN: A data mining concept. *Sustainable Computing: Informatics and Systems*, *23*, 120-135.

12. Bridges, S. M., & Vaughn, R. B. (2000). Intrusion detection via fuzzy data mining. In *12th Annual Canadian Information Technology Security Symposium*, 2(1), 109-122
13. Buczak, A. L., & Guven, E. (2015). A survey of data mining and machine learning methods for cyber security intrusion detection. *IEEE Communications surveys & tutorials*, *18*(2), 1153-1176.
14. Cannady, J. (2000, July). Applying CMAC-based online learning to intrusion detection. In *Proceedings of the IEEE-INNS-ENNS International Joint Conference on Neural Networks. IJCNN 2000. Neural Computing: New Challenges and Perspectives for the New Millennium*, 5, 405-410.
15. Cao, L. (2012). Social security and social welfare data mining: An overview. *IEEE Transactions on Systems, Man, and Cybernetics, Part C (Applications and Reviews)*, *42*(6), 837-853.
16. Chang, R. I., Lai, L. B., Su, W. D., Wang, J. C., & Kouh, J. S. (2007). Intrusion detection by backpropagation neural networks with sample-query and attribute-query. *International Journal of Computational Intelligence Research*, *3*(1), 6-10.
17. Conrad, M. (1988). Proton supermobility: a mechanism for coherent dynamic computing. *Journal of molecular electronics*, *4*(1), 57-65.
18. Desale, K. S., Kumathekar, C. N., & Chavan, A. P. (2015). Efficient intrusion detection system using stream data mining classification technique. In *2015 International Conference on Computing Communication Control and Automation*,1, 469-473.
19. Dewa, Z., & Maglaras, L. A. (2016). Data mining and intrusion detection systems. *International Journal of Advanced Computer Science and Applications*, *7*(1), 62-71.
20. Dhanabal, L., & Shantharajah, S. P. (2015). A study on NSL-KDD dataset for intrusion detection system based on classification algorithms. *International journal of advanced research in computer and communication engineering*, *4*(6), 446-452.
21. Eskin, E., Lee, W., & Stolfo, S. J. (2001). Modeling system calls for intrusion detection with dynamic window sizes. In *Proceedings DARPA Information Survivability Conference and Exposition II,* 1, 165-175.

22. Fan, W., Miller, M., Stolfo, S., Lee, W., & Chan, P. (2004). Using artificial anomalies to detect unknown and known network intrusions. *Knowledge and Information Systems*, *6*(5), 507-527.
23. Fathima, S., & Kiran, S. (2018). A Study on Network Security Administration using the Technology of Data Mining. *International Journal of Pure and Applied Mathematics*, *118*(24), 1-10.
24. Fekrazad, F. (2014). A best approach in intrusion detection for computer network PNN/GRNN/RBF. *International Journal of Computer Science Issues (IJCSI)*, *11*(1), 182.
25. Gaber, M. M., Aneiba, A., Basurra, S., Batty, O., Elmisery, A. M., Kovalchuk, Y., & Rehman, M. H. U. (2019). Internet of Things and data mining: From applications to techniques and systems. *Wiley Interdisciplinary Reviews: Data Mining and Knowledge Discovery*, *9*(3), 1292.
26. Ghosh, P., Ghosh, R., & Dutta, R. (2014). An alternative model of virtualization based intrusion detection system in cloud computing. *International Journal of Scientific & Technology Research*, *3*(5), 199-203.
27. Gómez, J., González, F., & Dasgupta, D. (2003). An immuno-fuzzy approach to anomaly detection. In *The 12th IEEE International Conference on Fuzzy Systems,* 2, 1219-1224.
28. Gu, C., & Zhang, X. (2009). A rough set and SVM based intrusion detection classifier. In *2009 Second International Workshop on Computer Science and Engineering*, 2, 106-110.
29. Hajimirzaei, B., & Navimipour, N. J. (2019). Intrusion detection for cloud computing using neural networks and artificial bee colony optimization algorithm. *ICT Express*, *5*(1), 56-59.
30. Hajisalem, V., & Babaie, S. (2018). A hybrid intrusion detection system based on ABC-AFS algorithm for misuse and anomaly detection. *Computer Networks*, *136*, 37-50.
31. Han, X. (2009, November). An improved intrusion detection system based on neural network. In *2009 IEEE International Conference on Intelligent Computing and Intelligent Systems*, 1, 887-890.
32. Hao, Y., Sheng, Y., & Wang, J. (2019). A graph representation learning algorithm for low-order proximity feature extraction to enhance unsupervised ids preprocessing. *Applied Sciences*, *9*(20), 4473.

33. Hossain, M., Bridges, S. M., & Vaughn, R. B. (2003). Adaptive intrusion detection with data mining. In *SMC'03 Conference Proceedings. 2003 IEEE International Conference on Systems, Man and Cybernetics. Conference Theme-System Security and Assurance,* 4, 3097-3103.
34. HU, L. N., & XU, W. B. (2007). Agent-based intrusion detection system using data mining approaches. *Computer Engineering and Design, 6, 5-10.*
35. Hwang, K., Cai, M., Chen, Y., & Qin, M. (2007). Hybrid intrusion detection with weighted signature generation over anomalous internet episodes. *IEEE Transactions on dependable and secure computing*, *4*(1), 41-55.
36. Ibrahim, L. M., Basheer, D. T., & Mahmod, M. S. (2013). A comparison study for intrusion database (Kdd99, Nsl-Kdd) based on self organization map (SOM) artificial neural network. *Journal of Engineering Science and Technology*, *8*(1), 107-119.
37. Idhammad, M., Afdel, K., & Belouch, M. (2018). Distributed intrusion detection system for cloud environments based on data mining techniques. *Procedia Computer Science*, *127*, 35-41.
38. Iman, A. N., & Ahmad, T. (2020). Data Reduction for Optimizing Feature Selection in Modeling Intrusion Detection System. *International Journal of Intelligent Engineering and Systems*, *13*(6), 199-207.
39. Jabez, J., & Muthukumar, B. (2015). Intrusion Detection System (IDS): Anomaly detection using outlier detection approach. *Procedia Computer Science*, *48*, 338-346.
40. Jianliang, M., Haikun, S., & Ling, B. (2009). The application on intrusion detection based on k-means cluster algorithm. In *2009 International Forum on Information Technology and Applications*, 1, 150-152.
41. Kandogan, E., & Haber, E. M. (2005). Security administration tools and practices. *Security and usability: Designing secure systems that people can use*, 4(2), 357-378.
42. Karimi, A. M., Niyaz, Q., Sun, W., Javaid, A. Y., & Devabhaktuni, V. K. (2016). Distributed network traffic feature extraction for a real-time IDS. In *2016 IEEE International Conference on Electro Information Technology, 4(3),* 522-526.

43. Kasongo, S. M., & Sun, Y. (2020). A deep learning method with wrapper based feature extraction for wireless intrusion detection system. *Computers & Security*, *92*, 101752.
44. Kim, G., Lee, S., & Kim, S. (2014). A novel hybrid intrusion detection method integrating anomaly detection with misuse detection. *Expert Systems with Applications*, *41*(4), 1690-1700.
45. Kirkpatrick, F. H. (1978). New models of cellular control: membrane cytoskeletons, membrane curvature potential, and possible interactions. *BioSystems*, *11*, 85-92.
46. Kshirsagar, V. K., Tidke, S. M., & Vishnu, S. (2012). Intrusion detection system using genetic algorithm and data mining: An overview. *International Journal of Computer Science and Informatics ISSN (PRINT)*, *2231*, 5292.
47. Lee, S. H., & Soh, J. (2003). Design and Implementation of the Intrusion Detection Pattern Algorithm Based on Data Mining. *The KIPS Transactions: PartC*, *10*(6), 717-726.
48. Lee, W., Fan, W., Miller, M., Stolfo, S. J., & Zadok, E. (2002). Toward cost-sensitive modeling for intrusion detection and response. *Journal of computer security*, *10*(1-2), 5-22.
49. Lee, W., Stolfo, S. J., & Mok, K. W. (2000). Adaptive intrusion detection: A data mining approach. *Artificial Intelligence Review*, *14*(6), 533-567.
50. Lee, W., Stolfo, S. J., Chan, P. K., Eskin, E., Fan, W., Miller, M., ... & Zhang, J. (2001). Real time data mining-based intrusion detection. In *Proceedings DARPA Information Survivability Conference and Exposition II,* 1, 89-100.
51. Li, D., Deng, L., Lee, M., & Wang, H. (2019). IoT data feature extraction and intrusion detection system for smart cities based on deep migration learning. *International journal of information management*, *49*, 533-545.
52. Liang, J., Lin, Q., Chen, J., & Zhu, Y. (2019). A filter model based on hidden generalized mixture transition distribution model for intrusion detection system in vehicle ad hoc networks. *IEEE Transactions on Intelligent Transportation Systems*, *21*(7), 2707-2722.
53. Lui, C. L., Fu, T. C., & Cheung, T. Y. (2005). Agent-based network intrusion detection system using data mining approaches. In *Third International Conference on Information Technology and Applications,* 1, 131-136.

54. Mohan Kumar, U., Siva SaiManikanta, P., & AntoPraveena, M. D. (2019). Intelligent security system for banking using Internet of Things. *Journal of Computational and Theoretical Nanoscience*, *16*(8), 3296-3299.

55. Molloy, I., Chen, H., Li, T., Wang, Q., Li, N., Bertino, E., ... & Lobo, J. (2010). Mining roles with multiple objectives. *ACM Transactions on Information and System Security (TISSEC)*, *13*(4), 1-35.

56. Mukhopadhyay, I., Chakraborty, M., Chakrabarti, S., & Chatterjee, T. (2011, December). Back propagation neural network approach to Intrusion Detection System. In *2011 International Conference on Recent Trends in Information Systems*, 1(2), 303-308.

57. Pan, S., Morris, T., & Adhikari, U. (2015). Developing a hybrid intrusion detection system using data mining for power systems. *IEEE Transactions on Smart Grid*, *6*(6), 3104-3113.

58. Panda, M., & Patra, M. R. (2008). A comparative study of data mining algorithms for network intrusion detection. In *2008 First International Conference on Emerging Trends in Engineering and Technology*, 4(6), 504-507.

59. Patel, R., Thakkar, A., & Ganatra, A. (2012). A survey and comparative analysis of data mining techniques for network intrusion detection systems. *International Journal of Soft Computing and Engineering (IJSCE)*, *2*(1), 265-260.

60. Patond, M. K., & Deshmukh, P. (2014). Survey on data mining techniques for intrusion detection system. *Int J Res Stud Sci Eng Technol*, *1*(1), 93-7.

61. Qi, B., & Dong, Y. F. (2012). A new model of intrusion detection based on data warehouse and data mining. In *Advanced Materials Research*, 383,v303-307.

62. Rajendran, P. K., Muthukumar, B., & Nagarajan, G. (2015). Hybrid intrusion detection system for private cloud: a systematic approach. *Procedia Computer Science*, *48*, 325-329.

63. Raut, R. G., & Gawali, S. Z. (2012). Intrusion detection system using data mining approach. *EXCEL International Journal of Multidisciplinary Management Studies*, *2*(8), 124-138.

64. Sabri, F. N. M., Norwawi, N. M., & Seman, K. (2011). Identifying false alarm rates for intrusion detection system with data mining. *IJCSNS:*

International Journal of Computer Science and Network Security, *11*(4), 95-99.

65. Sahin, Y., Bulkan, S., & Duman, E. (2013). A cost-sensitive decision tree approach for fraud detection. *Expert Systems with Applications*, *40*(15), 5916-5923.
66. Schadt, E. E., Li, C., Ellis, B., & Wong, W. H. (2001). Feature extraction and normalization algorithms for high-density oligonucleotide gene expression array data. *Journal of Cellular Biochemistry*, *84*(37), 120-125.
67. Singh, R. R., Gupta, N., & Kumar, S. (2011). To reduce the false alarm in intrusion detection system using self-organizing map. *International Journal of Soft Computing and Engineering (IJSCE)*, *1*(2), 4-19.
68. Singhal, A., & Jajodia, S. (2006). Data warehousing and data mining techniques for intrusion detection systems. *Distributed and Parallel Databases*, *20*(2), 149-166.
69. Smys, S., Basar, A., & Wang, H. (2020). Hybrid intrusion detection system for internet of Things (IoT). *Journal of ISMAC*, *2*(04), 190-199.
70. Solove, D. J. (2008). Data mining and the security-liberty debate. *The University of Chicago Law Review*, *75*(1), 343-362.
71. Stolfo, S. J., Lee, W., Chan, P. K., Fan, W., & Eskin, E. (2001). Data mining-based intrusion detectors: An overview of the columbia ids project. *ACM SIGMOD Record*, *30*(4), 5-14.
72. Strasburg, C., Stakhanova, N., Basu, S., & Wong, J. S. (2009). A framework for cost sensitive assessment of intrusion response selection. In *2009 33rd Annual IEEE international computer software and applications conference*, 1, 355-360.
73. Telikani, A., & Gandomi, A. H. (2019). Cost-sensitive stacked auto-encoders for intrusion detection in the Internet of Things. *Internet of Things*, 2(1), 100-122.
74. Thuraisingham, B., Khan, L., Masud, M. M., & Hamlen, K. W. (2008). Data mining for security applications. In *2008 IEEE/IFIP International Conference on Embedded and Ubiquitous Computing*, 2, 585-589.
75. Tian, Z. H., Fang, B. X., & Yun, X. C. (2003). An architecture for intrusion detection using honey pot. In *Proceedings of the 2003 International Conference on Machine Learning and Cybernetics,* 4, 2096-2100.

76. Vaidya, J., Atluri, V., & Guo, Q. (2010). The role mining problem: A formal perspective. *ACM Transactions on Information and System Security (TISSEC)*, *13*(3), 1-31.
77. Vaidya, J., Atluri, V., Warner, J., & Guo, Q. (2008). Role engineering via prioritized subset enumeration. *IEEE Transactions on Dependable and Secure Computing*, *7*(3), 300-314.
78. Wu, S. Y., & Yen, E. (2009). Data mining-based intrusion detectors. *Expert Systems with Applications*, *36*(3), 5605-5612.
79. Xiangrong, Y. A. N. G., & Junyi, S. Q. S. (2001). Data Mining Based Intelligent Intrusion Detection System [J]. *Computer Engineering*, *9*, 17-18.
80. Yan, B., & Han, G. (2018). Effective feature extraction via stacked sparse autoencoder to improve intrusion detection system. *IEEE Access*, *6*, 41238-41248.
81. Yan, K. Q., Wang, S. C., & Liu, C. W. (2009). A hybrid intrusion detection system of cluster-based wireless sensor networks. In *Proceedings of the International MultiConference of Engineers and Computer Scientists*, 1, 18-20.
82. Yong, H., & Feng, Z. X. (2010, November). Expert system based intrusion detection system. In *2010 3rd International Conference on Information Management, Innovation Management and Industrial Engineering*, 4, 404-407.
83. Yu, Y. (2010). A novel intrusion detection approaches based on data mining. In *2010 2nd International Conference on Computer Engineering and Technology*, 3, 349-351.
84. Yu, Z. X., Chen, J. R., & Zhu, T. Q. (2005). A novel adaptive intrusion detection system based on data mining. In *2005 International Conference on Machine Learning and Cybernetics*, 4, 2390-2395.
85. Zarrabi, A., & Zarrabi, A. (2012). Internet intrusion detection system service in a cloud. *International Journal of Computer Science Issues (IJCSI)*, *9*(5), 308.

Chapter 8

Recent Trends and Future Projections of Data Mining

CONTENTS

8.1. INTRODUCTION

As a novel research area, substantial progress has been made by data mining and has covered a wide spectrum of uses since the 1980s. Currently, data mining is utilized in an enormous array of areas. Several commercial data mining services and systems are accessible. Several challenges, though, remain. Here in this chapter, the mining of intricate data types as the introduction to additional in-depth study readers might select to do is introduced. Moreover, the trends and research limits in data mining are focused (Shaw et al., 2001; Ganguly et al., 2014).

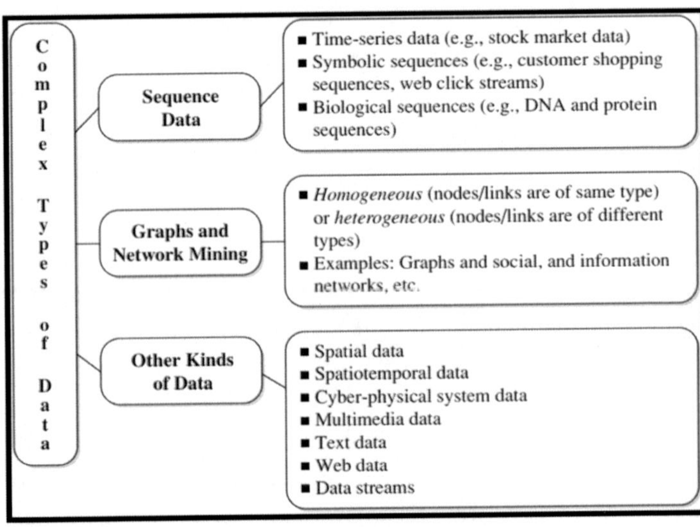

Figure 8.1. Intricate types of data for mining.

Source: https://studylib.net/doc/10135352/data-mining-trends-and-research-frontiers

8.2. A SEQUENCE OF DATA MINING: TIME-SERIES, SYMBOLIC AND BIOLOGICAL SEQUENCES

Sequences might be classified into 3 groups, centered on the features of the events they define (Lin et al., 2007):

- time-series data
- symbolic sequence data

- biological sequences

Time-series sequence data comprises of long series of numeric data, registered at equal intervals of time. The time-series data can generally be created by several economic and natural processes like stock markets, and medical, scientific, or natural observations.

Symbolic sequence data comprises of long series of nominal or event data, which usually aren't perceived at equal intervals of time. For various such kinds of sequences, *gaps* don't matter much. Instances comprise web clickstreams and customer shopping sequences, along with sequences of events in engineering and science and social and natural developments (Esling & Agon, 2012).

Biological sequences comprise protein and DNA sequences. Such kinds of sequences are usually very long and carry vital, complex, but secret semantic meaning. Gaps are normally very important here.

8.3. THE SEARCH OF RESEMBLANCE IN THE TIME-SERIES DATA

The set of time-series data comprises of series of numeric values attained over recurrent time measurements. The values are usually measured at equal intervals of time. Databases of time series are famous in numerous applications like analysis of the stock market, sales and economic projecting, budgetary examination, utility and inventory studies, yield and workload forecasts, and quality and process control. These databases are also beneficial for comprehending natural phenomena, engineering, and scientific experiments, and also for medical treatments (Damle & Yalcin, 2007).

Not like usual database queries, which locate data that match the given query precisely, a resemblance search discovers data sequences that fluctuate marginally from the given sequence of the query. Several time-series resemblance queries need subsequence matching, discovering the set of sequences that comprise subsequences that are analogous to the given sequence of a query (Serra & Arcos, 2014).

For resemblance search, it is frequently essential to first perform dimensionality or data reduction and conversion of the time-series data. Usual dimensionality reduction methods include:

- DFT (discrete Fourier transform)

- DWT (discrete wavelet transform)
- SVD (singular value decomposition) centered on PCA (principal components analysis).

These characteristics form the characteristic space, which is the projection of converted space. Indices can normally be made on the converted or original time-series data in order to speed up the search. For the resemblance search based on a query, techniques comprise normalization transformation, window stitching, subsequence ordering, and atomic matching (Faloutsos et al., 1994). Various software packages are available for resemblance search in the time-series data.

8.4. ANALYSIS OF REGRESSION AND TREND IN THE TIME-SERIES DATA

Regression examination of the time-series data has generally been studied considerably in the areas of signal and statistics analysis. Though, one might frequently go beyond the pure analysis of regression and execute trend analysis for various practical uses (Bernal et al., 2017).

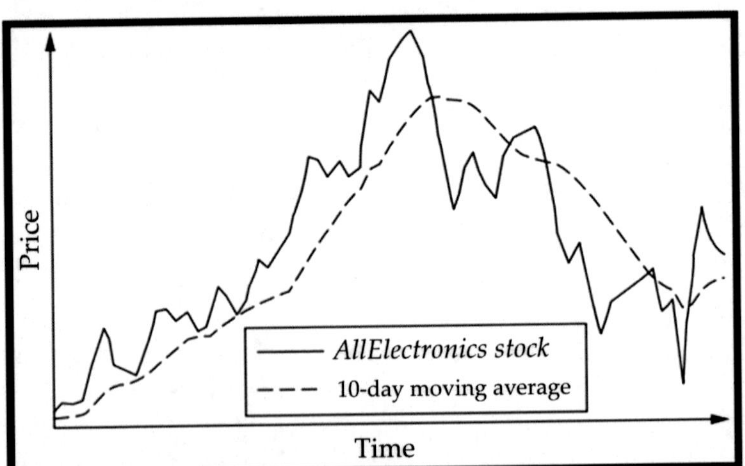

Figure 8.2. The database of time-series for the cost of all of the electronics over time. The drift is displayed with the dashed curve, computed by the moving average.

Source: https://www.semanticscholar.org/paper/Data-Mining-Trends-and-Research-Frontiers-Han-Kamber/dd84d770c620b3bf71615dcd22230036004dfe96

Analysis of trend makes a combined model utilizing the following 4 main movements or components to describe the time-series data (Visser & Molenaar, 1995; Bhaskaran et al., 2013):

- Long-term movements: These movements specify the overall direction in which the graph of time-series is going overtime, for instance, utilizing the least squares and weighted moving average methods in order to locate trend curves.
- Cyclic movements: These movements are long-term oscillations regarding the curve or a trend line.
- Seasonal variations: These discrepancies are almost indistinguishable patterns that the time series seems to follow throughout consistent seasons of consecutive years like holiday shopping seasons. For efficient analysis of trends, the data frequently must be deseasonalized centered on the seasonal index calculated by autocorrelation.
- Random movements: These movements describe sporadic fluctuations because of chance events like labor quarrels or declared personnel variations within companies.

The analysis of trends can also be utilized for time-series projecting, finding the mathematical function that will nearly produce the historic patterns in the time series, and utilizing it to create short-term or long-term forecasts of future values.

8.5. ADVANCEMENT OF INFORMATION AND SOCIAL NETWORKS

Networks are vibrant and continuously developing. Identifying developing communities and developing regularities or abnormalities in heterogeneous or homogeneous networks can aid people in better comprehend the structural advancement of networks and forecast irregularities and trends in developing networks (Combs, 2003). For the homogeneous networks, the progressing communities found are sub-networks comprising of objects of similar type like the set of coauthors or friends. Conversely, for the heterogeneous networks, the advancing communities found are sub-networks comprising of objects of dissimilar types, like the associated set of authors, papers, terms, and venues, from which one can derive the set of developing objects for each kind, like progressing themes and authors.

8.6. DATA MINING USES

The principles and techniques for mining data warehouses, relational data, and intricate data types have been described in this book. Since data mining is a comparatively new discipline with broad and diverse uses, there still exists a nontrivial gap amongst overall principles of mining of data and particular application-based, efficient data mining tools (Setoguchi et al., 2008).

8.6.1. Mining of Data for Financial Data Examination

Most financial institutions and banks offer a broad variety of investment, banking, and credit services. Some also provide services of insurance and investment in stock.

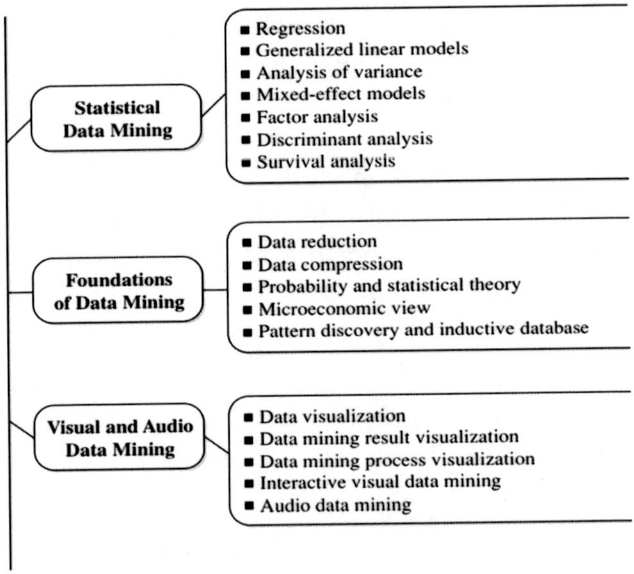

Figure 8.3. Typical application domains of data mining.

Source: https://studylib.net/doc/10135352/data-mining-trends-and-research-frontiers

Financial data gathered in the financial and banking industry are frequently comparatively complete, trustworthy, and of quality, which assists systematic data scrutiny and data mining. Some of the usual cases are presented here (Kirkos et al., 2007).

8.6.1.1. Design and building of data warehouses for intricate data examination and data mining

Like several other uses, data warehouses must be made for financial and banking data. Multidimensional data examination techniques should be utilized to examine the overall properties of such kinds of data (D'Oca & Hong, 2015). For instance, the financial officer of a company might want to look at debt and revenue fluctuations by region, sector, and month and some other factors, as well as minimum, maximum average, deviancy, trend, and some other statistical data. Data cubes, data warehouses, categorization and class evaluations, and outlier scrutiny will all play significant roles in data scrutiny and mining.

8.6.1.2. Loan payment forecast and the customer credit policy examination

Forecast of the loan payment and customer credit examination is crucial to the bank business. Various factors can weakly or strongly affect the performance of loan payment and the rating of customer credit (Pike & Cheng, 2001). Methods of data mining, like attribute relevance ranking and attribute selection, might help recognize significant factors and eradicate unrelated ones. For instance, factors associated with the hazard of loan payments comprise loan to value ratio, debt ratio, term of the loan, income level of customer, education level, region of residence, and history of credit. Examination of payment history of a customer might discover that, say, payment to income ratio is the leading factor, whereas debt ratio and education level aren't. The bank might then decide to amend its policy of granting a loan to those customers/clients whose applications were rejected previously but whose profiles exhibit comparatively low jeopardies according to the crucial factor analysis (Hurley et al., 2019).

8.6.1.3. Cataloging and grouping of clients for targeted marketing

Cataloging and grouping methods can be utilized for client group recognition and targeted marketing. For instance, one can utilize cataloging to recognize the most critical factors that might affect the decision of a customer regarding banking. Clients with the same behaviors regarding payments of loans might be recognized by multidimensional grouping techniques. These can aid recognize customer groups, relate the new customer with a suitable customer group and assist targeted marketing (Linden et al., 2003).

8.6.1.4. Revealing money laundering and various other financial crimes

In order to reveal money laundering and various other financial misconducts, it is vital to assimilate information from various, heterogeneous databases, as long as these databases are potentially associated with the study. Various tools of data analysis can then be utilized to perceive unusual patterns, like cash flow patterns at particular periods, by particular groups of customers. Helpful tools comprise data visualization tools, association and information network examination tools, cataloging tools, grouping tools, outlier examination tools, and chronological pattern examination tools. These tools might recognize significant associations and activities patterns and assist investigators to focus on doubtful cases for more detailed analysis (Gonzalez, 1996; Cuéllar, 2002).

8.6.2. Mining of Data for Telecommunication and Retail Industries

The retail business is a suitable application field for mining data, as it gathers large amounts of data on the shopping history of a customer, goods transportation, sales, service, and consumption. The amount of data gathered continues to expand quickly, especially because of the increasing accessibility, ease, and reputation of business shown on the Web. Currently, most of the main chain stores have websites where customers can shop online (Ramageri & Desai, 2013). Some of the businesses, like Amazon.com, exist only online. Retail data offer a rich source for the mining of data.

Retail mining of data can aid recognize buying behaviors of clients, discover shopping patterns and drifts of customers, enhance the customer service quality, accomplish better customer maintenance and satisfaction, improve ratios of goods consumption, design more efficient goods carriage and supply policies, and decrease the expense of a business. Some instances related to the mining of data in the retail business are given as follows (Hormozi & Giles, 2004):

8.6.2.1. Design and building of data warehouses

Since retail data cover the broad spectrum (comprising sales, employees, customers, transportation of goods, services, and consumption), there usually can be several ways to design the data warehouse for this particular industry. The detail levels to comprise can vary considerably. The consequence of initial data mining drills can be utilized to assist help the design and expansion of data warehouse structures. This includes deciding which levels and dimensions to comprise and what preprocessing to execute to enable efficient mining of data (Golfarelli et al., 1998).

8.6.2.2. Multidimensional examination of sales, products, clients, region, and time

The retail business needs well-timed information regarding the needs of the customer, sales of products, future trends, and styles, along with the expense, profit, quality, and service of supplies. It is thus significant to offer powerful multidimensional examination and visualization tools, comprising the creation of advanced data cubes conforming to requirements of data examination (Bachlechner et al., 2014). The *advanced data cube structures* are helpful in retail data examination as they aid examination on multidimensional groups with intricate conditions.

8.6.2.3. Examination of the efficiency of sales promotions

The retail business conducts sales promotions utilizing coupons, advertisements, and numerous kinds of bonuses and discounts in order to stimulate products and appeal to customers. Careful examinations of the efficiency of sales promotions can aid enhance the profits of a company. A multidimensional examination can be utilized for this drive by associating the sales and transactions comprising the items of sales during the period of sales versus those comprising similar items after or before the sales promotion (Twedt, 1952). Furthermore, association examination might reveal which items are probable to be bought together with the sale items, particularly in contrast with the sales after or before the promotion.

8.6.2.4. Customer retaining-examination of client loyalty

One can utilize the information of customer loyalty cards to record a series of purchases of specific customers. The loyalty of customers and trends of purchase can be examined systematically. The items purchased at diverse periods by the particular set of customers can be clustered into sequences. The sequential pattern mining of data can then be utilized to investigate variations in customer loyalty or consumption and recommend variations on the variety and pricing of items to aid retain customers and invite new ones (Eggert, 2011).

8.6.2.5. Endorsement of products and cross-referencing of goods

By mining relations from sales registries, one might find that the customer who purchases the digital camera is probable to purchase another set of goods. Such kind of information can be utilized to develop product recommendations. The collaborative recommender systems utilize data mining methods to make custom product recommendations throughout live client transactions, centered on the thoughts of some other customers (Thies et al., 2019). The recommendations of products can be promoted on sales receipts or on the Web site to aid enhance customer service, help customers in choosing goods, and ultimately increasing sales. Likewise, information, like hot goods this week or eye-catching deals, can be shown together with the information in order to encourage sales.

8.6.2.6. Fraudulent examination and the recognition of strange patterns

Fraudulent activity charges the retail business millions of dollars each year. It is vital to (a) recognize possibly fraudulent customers and their unusual patterns of usage; (b) perceive efforts to attain fraudulent entry or illegal access to organizational and individual accounts, and (c) find strange patterns that might require special attention. Various patterns can be found by multidimensional examination, cluster examination, and outlier examination (Ngai et al., 2011).

As one more industry that manages large amounts of data, the industry of **telecommunication** has rapidly developed from providing local and global telephone services to offering several other complete communication services. These comprise cellular phones, Internet access, smartphone, text messages, email, pictures, and web data broadcasts. The incorporation of telecommunication, Internet, computer network, and several other means of computing and communication has been on way, altering the face of computing and telecommunications (Ravisankar et al., 2011). This has made the great demand for the mining of data in order to aid comprehend business dynamics, recognize telecommunication patterns, identify false activities, make better utilization of resources, and enhance the quality of service.

8.6.3. Mining of Data in Engineering and Science

In history, various scientific data examination tasks tended to manage comparatively small and homogeneous sets of data. Such types of data were usually analyzed using the *formulate hypothesis, make model, and assess outcomes*" paradigm. In these circumstances, statistical methods were usually engaged for their examination (Marbán et al., 2009). Huge data gathering and data storage technologies have currently altered the background of scientific data examination. Nowadays, scientific data can generally be combined at lower costs and much higher speeds. This has occasioned the gathering of enormous volumes of stream data, high-dimensional data, and heterogeneous data, comprising ample spatial and sequential information. Therefore, scientific uses are fluctuating from the hypothesize and test model towards the gather and store data, data mining for novel hypotheses, validate with data or experimentation procedure (Hossain et al., 2019).

8.6.3.1. Preprocessing of data and data warehouses

Preprocessing of data and data warehouses are crucial for the exchange of information and mining of data. Making the warehouse frequently needs discovering means for solving unpredictable or unsuited data gathered in several environments and at diverse periods. This needs reconciling semantics, systems of referencing, geometry, correctness, measurements, and precision. Techniques are required for incorporating data from heterogeneous sources and also for recognizing events (Orlando et al., 2007).

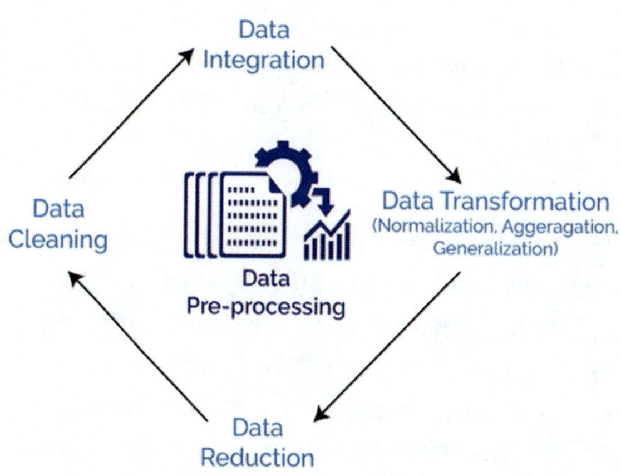

Figure 8.4. Pre-processing cycle of data

Source: https://www.electronicsmedia.info/2017/12/20/what-is-data-prepro-cessing/

For example, consider ecosystem and weather data, which are chronological and spatial and need cross-referencing geospatial data. The main issue in examining such kind of data is that there exist numerous events in the spatial domain but only a few in a chronological domain. For instance, El Nino events happen only every 4 to 7 years, and former data might haven't been gathered as methodically as they are gathered today (Dietrich et al., 2018). Techniques are also required for the effective calculation of intricate spatial aggregates and the management of spatial-associated data streams.

8.6.3.2. Mining intricate data types

The sets of scientific data are usually heterogeneous. They normally include unstructured and semi-structured data, like geo-referenced stream data and multimedia data, along with data having intricate, hidden semantics. Vigorous and devoted analysis approaches are required for managing biological data, associated theory hierarchies, and intricate semantic associations. For instance, in the area of bioinformatics, the research issue is to recognize regulatory impacts on genes. The *gene regulation* mentions how genes in the cell are switched off (or on) to define the functions of a cell. Diverse biological processes include diverse sets of genes performing together inaccurately regulated patterns (Liu et al., 2016). Therefore, to comprehend the biological process one needs to recognize the contributing genes and the regulators. This needs the development of intricate data mining techniques to examine large sets of biological data for clues regarding regulatory impacts on particular genes, by discovering DNA segments facilitating such impact.

8.6.3.3. Graph-centered and network-centered mining

It is frequently problematic or unmanageable to model various physical phenomena and procedures because of the limitations of prevailing modeling methods. Instead, labeled networks and graphs might be utilized to capture several spatial, biological, geometric, and some other relational features existent in the sets of scientific data. In network or graph modeling, every object which is to be mined is characterized by the vertex in a graph, and the edges between vertices signify associations between objects (Pappalardo et al., 2016). For instance, these graphs can be utilized to classic chemical structures, biological paths, and data produced by numeric simulations like fluid-flow simulations. The accomplishment of network or graph modeling, though, relies on enhancements in the efficiency and scalability of numerous graph-centered data mining tasks like classification, regular pattern mining, and grouping.

8.6.3.4. Tools for visualization and knowledge of a particular domain

High-level GUIs (graphical user interfaces) and tools of visualization are needed for systems of scientific data mining. These must be incorporated with prevailing particular domain data and the information systems in order to assist researchers and overall users in examining patterns, understanding and visualizing found patterns, and utilizing knowledge in the decision making (Goebel & Gruenwald, 1999).

Data mining in the area of engineering has many resemblances with the mining of data in science. Both of the practices frequently gather huge amounts of data and need data preprocessing and warehousing, and accessible mining of intricate data types. Both normally utilize visualization and make good utilization of networks and graphs. Furthermore, several engineering processes require real-time reactions, and thus mining data streams in real-time frequently becomes a crucial component (Chen et al., 2008).

Huge amounts of communication data dispense into daily life. Such kind of communication data prevails in several forms, comprising news, articles, online discussions, twitters, reviews of products, messages, and communications, on the Web and in several types of social networks. Therefore, the mining of data in social studies and social science has become gradually popular. Furthermore, feedback of users regarding speeches, articles, and products can be examined to gather overall views and sentiments. The results of an analysis can be utilized to forecast trends, enhance work, and aid in the making of a decision (Cannataro et al., 2004).

Computer science produces exclusive data kinds. For instance, computer programs can often be long, and the execution of these programs frequently creates huge-size traces. Computer networks can generally have intricate structures and the flows of a network can be vibrant and huge. Sensor networks might produce huge amounts of data with diverse dependability. Computer databases and systems can experience several types of attacks, and the system might raise safekeeping and privacy concerns. The exclusive data kinds offer rich land for the mining of data.

Mining of data in computer science can generally be utilized to assist the status of the monitor system, enhance the performance of a system, isolate bugs of software, notice software bootlegging, examine faults of a computer system, expose network interruptions, and identify system faults (Shneiderman, 2002). Mining of data for system engineering and software can function on dynamic or static data, reliant on either the system dumps traces before for post-examination or if it should react in real-time in order to manage online data.

8.6.4. Mining of data and Recommended Systems

Consumers nowadays are confronted with numerous services and goods when purchasing online. Recommender systems assist a customer by making product references that are probable to be of concern like books, movies, online news articles, restaurants, and some other services. The recommender systems might utilize the content-centered method, the collaborative method, or the hybrid method that merges both of the above methods (Najafabadi et al., 2019).

The content-centered method endorses items that are analogous to items the customer preferred in the past. It depends on product characteristics and textual item explanations. The collaborative method might consider the social environment of a user. It mentions items centered on the views of other users who have analogous preferences as the customer. The recommender systems utilize a wide range of methods from retrieval of information, statistics, and mining of data to search for resemblances among goods and preferences of the customer.

The benefit of these recommender systems is that these systems offer personalization for users of e-commerce, encouraging one to one marketing. Amazon, the innovator in the utilization of collaborative recommender systems, provides "the personalized store for each client" as part and parcel of their marketing approach (Moreno et al., 2016). Personalization can help both user and the company. By having more precise models of their clients, companies gain a better understanding of the needs of the customer. Attending these needs can outcome in greater success in cross-selling of associated products, product affinities, upselling, larger baskets, and the retention of the customer.

The recommendation issue contemplates the set of users, C, and the set of items, S. Let u be the utility function that computes the worth of the item, s, to the user, c. The utility is normally characterized by the rating and is primarily described only for goods rated formerly by users. For instance, when joining the system of movie recommendations, users are usually asked to rate various movies. Space $S \times C$ of all probable users and goods is enormous (Park et al., 2012). The recommendation system must be able to infer from known to unfamiliar ratings in order to forecast item–user amalgamations. Goods with the highest prophesied rating for the user are suggested to that user.

The collaborative recommender system attempts to forecast the utility of goods for the user, *u*, centered on items formerly rated by some other customers who are analogous to *u*. For instance, when endorsing books, the collaborative recommender system efforts to find other customers who have a history of approving with *u*. The collaborative recommender systems can be memory-based or centered on a model (Nilashi, 2016).

Memory-based approaches fundamentally utilize heuristics in order to make rating forecasts centered on the complete collection of items rated formerly by customers. The unfamiliar rating of the item–user amalgamation can be assessed as the collection of ratings of most comparable customers for the same product. Usually, the *k*-nearest-neighbor method is utilized, that is, one finds the *k* other customers that are most comparable to the target customer, *u*. Several strategies can be utilized to calculate the similarity amongst customers. The most famous method uses either cosine similarity or Pearson's correlation coefficient (Kim et al., 2010). The weighted aggregate can generally be utilized, which alters for the point that different customers might utilize the rating scale contrarily. Model-centered collaborative recommender systems utilize the group of ratings in order to learn the model, which is utilized to make rating forecasts. For instance, grouping, Bayesian networks, probabilistic models, and some other machine learning methods have been used.

Recommender systems suffer main challenges like scalability and confirming quality endorsements to the user. For instance, about scalability, the collaborative recommender systems should be able to find through millions of possible neighbors in real-time. If a site is utilizing browsing patterns as signs of item preference, it might include numerous data points for its users (Park et al., 2011). Confirming quality recommendations is critical to gain the trust of the consumer. If users follow the system recommendation but don't like the item, they are less probable to use this recommender system again.

As with the systems of classification, the recommender systems can usually make 2 kinds of errors: false positives and false negatives. F*alse negatives* are items or goods that the system flops to endorse, even though the user would like these items. *False positives* are items or goods that are suggested, but which the user doesn't like. Content-centered recommender systems are restricted by the characteristics utilized to define the items they suggest (Gholamian et al., 2011). Another test for content-centered and collaborative recommender systems is to deal with novel customers for which the purchasing history isn't yet accessible.

Hybrid strategies incorporate both content-centered and collaborative approaches to accomplish further enhanced endorsements. The Netflix Award was a competition held by the DVD-rental service online, with the payout of 1,000 thousand dollars for the finest recommender algorithm in order to forecast customer ratings for movies, centered on former ratings (Da Costa & Manzato, 2016). This competition and some other studies have exhibited that the projecting exactness of the recommender system can generally be considerably enhanced when blending several predictors, particularly by utilizing an ensemble of numerous considerably different approaches, instead of refining the single method.

8.7. MINING OF DATA AND SOCIETY

For the majority of people, mining data is part of their daily lives, even though one might often be ignorant of its existence. This section highlights various instances of ubiquitous and invisible mining of data, affecting things of daily routine from the items stocked at the supermarket, to the advertisements one sees while browsing the Internet, to the prevention of crime. Data mining can provide the individual various benefits by enhancing customer service and gratification along with lifestyle. Though, it has serious consequences about one's right to security and privacy of data (Romero & Ventura, 2013).

8.7.1. Ubiquitous and Invisible Mining of Data

Data mining is existent in various aspects of one's daily life. It disturbs how one shops, works, and searches for information and can also impact leisure time, well-being, and health. In this segment, instances of such ubiquitous mining of data are looked at. Many of these instances also signify invisible mining of data, in which smart software, like customer-adaptive web services, search engines, intelligent database systems, and ticket masters, integrates mining of data into its operational components, frequently unknown to the customer (Borriello, 2000).

From the grocery supermarkets that print out coupons on the receipts of customers to online shopping stores that endorse additional goods centered on user interests, mining of data has creatively influenced what one buys, and one's experience while purchasing. One instance is Wal-Mart, which has billions of users visiting its stores each week. Wal-Mart permits suppliers to access information on their items and perform examinations utilizing the software of data mining (Mattern et al., 2001). This permits suppliers to recognize user buying patterns at diverse stores, control the record and placement of products, and recognize novel merchandising chances. All these factors affect which products end up on the shelves of stores.

The mining of data has changed the online purchasing experience. Many customers regularly turn to online shopping stores in order to buy books, movies, and music (Borriello, 2008). Recommender systems provide custom-made product recommendations centered on the views of some other customers. Amazon was at the front of utilizing such a tailored, data mining–centered method as the strategy of marketing. It has been perceived that in customary brick and mortar chain, the most difficult part is getting the client into the chain. Once the user is there, he/she is expected to purchase something, as the price of going towards another outlet is high. Thus, the marketing strategy for brick and mortar stores inclines to stress drawing users in, instead of the real in-store user experience. This is in divergence from online shopping stores, where users can walk out and check in to another store with only a mouse click (Beer, 2007; Borgavakar & Shrivastava, 2017).

Various companies increasingly utilize mining of data for CRM (customer relationship management), which aids offer more tailored, personal service attending to the needs of an individual customer, instead of mass marketing. By understanding browsing and shopping patterns on online stores, the companies can alter advertisements and campaigns to profiles of users, so that users are less probable to be irritated with undesirable mass mailings. These activities can outcome in considerable expense savings for companies. The users further get an advantage that they are likely to be informed of deals that are really of interest, occasioning in less wastage of time and greater gratification (Gane et al., 2007).

Data mining has affected the manners in which individuals use computers and look for information. Once the Internet is connected, for instance, one decides to check his/her email. Unknown to the user, several irritating emails have been deleted already, thanks to the spam filter that utilizes classification algorithms to identify spam. After processing the email, the user goes to Google, which offers access to information from numerous web pages indexed on the server. Google is amongst the most famous and broadly utilized Internet search engines (Sarasvathi & Fong, 2008). Utilizing Google to find information has become the way of life for several people.

Google is so famous that it has become the novel verb in the English language. The user decides to type in some words for the topic of concern. Google returns the list of web pages on the topic, organized by the set of mining of data algorithms comprising PageRank. Furthermore, if one types Boston New York, Google will display the train and bus timetables from Boston to New York; though, the slight change to Boston Paris will display the flight timetables from Boston to Paris (Dimitoglou et al., 1998).

While one is observing the results of a Google query, several advertisements pop up linking to the query. The strategy of Google of altering ads to match the interests of users is one of the usual services being discovered by almost every Internet search provider.

Mining of data is universal, as can generally be observed from the daily-faced instances. In most circumstances, mining of data is invisible, as customers might be ignorant that they are assessing outcomes returned by the mining of data or that the mouse clicks are fed as novel data into some of the data mining functions (Indra, 2007). For the mining of data to become further enhanced and accepted as the technology, enduring research and development are required in the various fields mentioned as challenges all over this book. These comprise effectiveness and scalability, augmented user interaction, integration of background knowledge and the methods of visualization, effective approaches for discovering interesting patterns, enhanced management of intricate types of data and stream data, web mining, and real-time data mining. Additionally, the *incorporation* of mining of data into prevailing business and specific scientific technologies, to offer data mining tools of a particular domain, will further back to the development of technology (Bailey & Urquhart, 2003). The accomplishment of data mining clarifications personalized for e-commerce uses, as conflicting to general systems of data mining, is an instance.

8.7.2. Privacy, Safety, and Social Influences of Data Mining

With more information available in electronic forms and also available on Web pages, and with progressively powerful tools of data mining being made and put into usage, there are growing worries that data mining might pose the risk to the privacy and security of data. Though, it is significant to observe that several data mining applications don't even touch private data. Famous instances include applications comprising natural resources, forecast of droughts and floods, meteorology, geography, biology, and some other engineering and scientific data. Moreover, most of the studies in research of data mining focus on the advancement of scalable algorithms and don't comprise personal data (Allam et al., 2006).

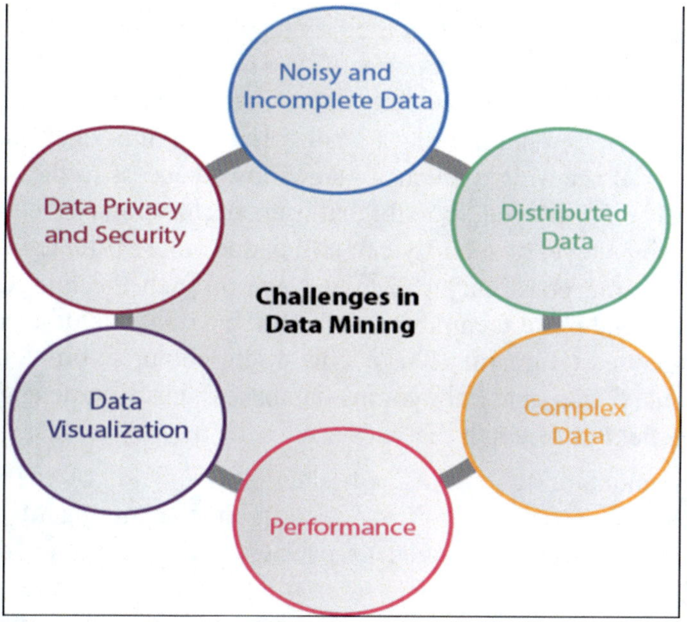

Figure 8.5. Challenges faced in systems of data mining

Source: https://www.javatpoint.com/data-mining

The emphasis of data mining is on the *finding of generic or statistically important patterns* and not on particular information about individuals. It is believed in this sense that the actual privacy anxieties are with free access to individual accounts, particularly access to sensitive information like transaction records of credit cards, personal financial records, criminal investigations, and also ethnicity (Xu et al., 2014). For the applications of data mining that do comprise personal data, in various circumstances, simple techniques like eradicating sensitive IDs from data might safeguard the privacy of individuals. However, privacy anxieties prevail wherever personally recognizable information is gathered and kept in digital form and the programs of data mining are capable to access such kind of data, even throughout the preparation of data.

Improper or absent disclosure control can generally be the main cause of privacy concerns. To manage such issues, various data security-improving techniques have been made. Additionally, there exists a great deal of current effort on advancing *privacy maintaining* data mining techniques (Bryce & Klang, 2009).

"What can be done to protect the privacy of persons whereas gathering and mining data?" Various data security–improving methods have been developed in order to help secure data. The databases can engage the multilevel security model in order to categorize and limit data according to several levels of security, with customers allowed access to their permitted level. It has been displayed, though, that users performing particular queries at the authorized level of security can still deduce more private information and that the same possibility can take place through the mining of data. Encryption is one more technique in which the data items of an individual might be encoded (Bignami, 2007). This might comprise blind signatures, biometric encryption, and anonymous databases. Intrusion detection is one more active field of research that aids secure the privacy of data.

Privacy-maintaining mining of data is a field of data mining investigation in reaction to privacy security in the mining of data. It is also called the privacy-improved or privacy-sensitive mining of data. It deals with attaining effective data mining outcomes without revealing the fundamental values of sensitive data. Most of the privacy-maintaining data mining techniques utilize some kind of alteration on data in order to execute privacy preservation. Usually, such techniques decrease the granularity of demonstration to maintain privacy. For instance, they might generalize the information from the individual user to groups of users (Dixit & Pandya, 2014). This granularity reduction triggers information loss and probably the worth of data mining outcomes. This is the trade-off between privacy and information loss. Privacy-maintaining data mining techniques can be categorized into the following classes.

8.7.2.1. Randomization techniques

These approaches add some noise to data in order to mask some quality values of records. The added noise should be adequately large enough so that the record values of an individual, particularly sensitive ones, can't be recovered. Though, it must be added competently so that the ultimate outcomes of mining data are fundamentally preserved. Methods are developed to derive total distributions from the disturbing data. Consequently, data mining methods can be made to work with aggregate distributions (Kallio et al., 2011).

8.7.2.2. The methods of l-diversity and k-anonymity

These methods change the records of individuals so that they can't be exclusively recognized. In the technique of *k-anonymity*, the granularity of representation of data is decreased adequately so that any record maps onto as a minimum *k* other records in data. It utilizes methods such as suppression and generalization. The method of *k*-anonymity is weak, if values are homogeneous within the group, then these values might be concluded for the changed records (Rajendran et al., 2017). The model of *l-diversity* was made to manage this weakness by imposing an intragroup variety of sensitive values in order to confirm anonymization. The objective is to make it adequately hard for adversaries to utilize amalgamations of record characteristics to exactly recognize individual records.

8.7.2.3. Distributed privacy maintenance

Large sets of data could be segregated and disseminated either *horizontally* or *vertically*, or even in an amalgamation of both. Whereas the individual sites might not want to distribute their complete sets of data, they might consent to restricted sharing of information with the utilization of the range of protocols. The general effect of such kinds of methods is to preserve privacy for every individual object, whereas deriving aggregate outcomes overall data (Gunay & Shen, 2017).

8.7.2.4. Downgrading the efficiency of data mining outcomes

In numerous cases, although the data might not be accessible, the yield of data mining might outcome in defilements of privacy.

Currently, researchers proposed novel concepts in privacy-maintaining data mining like the idea of differential privacy. The overall idea is, for any 2 sets of data that are very close to each other, the given differentially private algorithm will act nearly the same on both of the sets of data. This definition provides a strong assurance that the existence or lack of the small data set won't disturb the ultimate output of the query considerably (Vaghashia & Ganatra, 2015). Centered on this idea, the set of differential privacy-maintaining data mining algorithms have usually been developed. Exploration in this field is ongoing. More powerful privacy-maintaining data publishing and mining of data algorithms are expected in near future.

Similar to other technologies, the mining of data can be misused. Though, one must not lose the prospect of all the advantages that research of data mining can bring, varying from insights obtained from medical applications to augmented customer gratification by aiding companies better match their clients' requirements (Verykios et al., 2004). It is expected that computer scientists, counter-terrorism experts, and also policy experts will carry on to work with lawyers, social scientists, companies, and users to take charge in building ways to confirm data privacy safety and protection. In this manner, one might continue to gain the advantages of mining data in terms of money and time savings and the innovation of novel knowledge.

8.8. TRENDS OF DATA MINING

The multiplicity of data, tasks of data mining, and the data mining methods pose various challenging research problems in data mining. The advancement of effective and efficient data mining approaches, services and systems, and interactive and incorporated environments of data mining is the main area of study. The utilization of data mining methods to solve complex or large application issues is a significant job for data mining systems and data mining researchers and application designers (Kriegel et al., 2007). This section defines some trends in the mining of data that imitate the chase of these challenges.

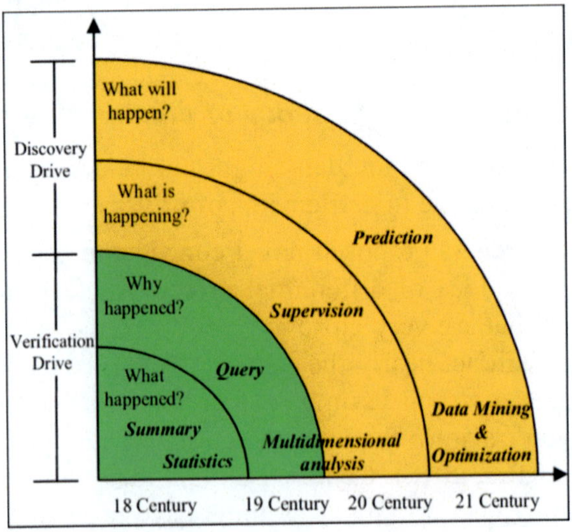

Figure 8.6. Mining of data and data examination trends

Source: https://www.researchgate.net/figure/Trends-of-Data-Analysis_fig1_275893871

8.8.1. Application assessment

Early applications of data mining put many efforts into aiding businesses to gain a competitive edge. The assessment of mining of data for businesses remains to magnify as e-marketing and e-commerce have become normal in the retail business. Data mining is gradually utilized for the assessment of applications in areas like text and web analysis, industry, biomedicine, science, and government. Evolving application areas comprise the mining of data for counter-terrorism and mobile data mining (Mandala et al., 2012). Since general data mining systems might have restrictions in dealing with issues of a particular application, one might see the trend towards the advancement of more particular application data mining tools and systems, along with invisible data mining tasks fixed in several kinds of services.

8.8.2. Scalable and collaborating methods of data mining

In comparison with customary data analysis approaches, data mining should be capable to manage large amounts of data proficiently and, if probable, interactively. Since the quantity of data being gathered continues to upsurge quickly, scalable algorithms for incorporated and individual data mining tasks become necessary (Koren, 2010). One significant direction towards enhancing the general efficiency of the process of mining whereas increasing customer interaction is constraint-centered mining. This offers users extra control by permitting the specification and utilization of constraints in order to assist systems of data mining in the search for fascinating patterns and knowledge.

8.8.3. Incorporation of mining of data with database systems, search engines, systems of data warehouse, and systems of cloud computing

Search engines, systems of a database, systems of data warehouse, and systems of cloud computing are typical computing and information processing systems. It is significant to confirm that data mining acts as a necessary data examination component that can normally be smoothly incorporated into such kind of information processing environment (Wu et al., 2013). The data mining service must be firmly coupled with such kinds of systems as the seamless, integrated framework or as the invisible function. This will confirm the availability of data, data mining compactness, high performance, and the incorporated environment of information processing for multidimensional data examination and exploration.

8.8.4. Mining information and social networks

Mining information and social networks and link examination are crucial tasks as such kinds of networks are intricate and omnipresent. The advancement of scalable and efficient knowledge discovery techniques and applications for huge numbers of networks data is necessary (Irfan et al., 2015).

8.8.5. Mining moving objects, Spatio-temporal, and cyber-physical systems

Cyberphysical systems along with spatiotemporal data are increasing quickly because of the famous utilization of cellular phones, sensors, and some other wireless equipment. Numerous challenging research problems are realizing real-time and efficient knowledge innovation with such kind of data (Chen et al., 2017).

8.8.6. Mining web, multimedia, and text data

Mining such data kinds is a current emphasis in the research of data mining. Great advancement has already been made, still, there are numerous open problems to be answered (Zaiane et al., 1998).

8.8.7. Mining biomedical and biological data

The exclusive amalgamation of intricacy, abundance, size, and significance of biomedical and biological data permits special consideration in the mining of data. Mining protein and DNA sequences, mining biological paths, and network examination are some topics in this area. Other fields of biological research of data mining comprise mining biomedical literature and information incorporation of biological data with the help of data mining (Rosania et al., 2007).

8.8.8. Mining of data with system and software engineering

The large computer systems and software programs have become gradually bulky in size refined in intricacy, and incline to originate from the incorporation of various components advanced by diverse execution teams. This inclination has made it a progressively challenging job to confirm software robustness and dependability. The examination of the implementations of the software program with bugs is the process of data mining—finding the data produced during program implementations might disclose significant outliers and patterns that might lead to ultimate automated innovation of software bugs (Xie et al., 2009). It is expected that further advancement of data mining practices for software debugging will improve the robustness of software and bring novel vigor to software and system engineering.

8.8.9. Audio and visual data mining

Audio and visual mining of data is an efficient way to incorporate with humans' audio and visual systems and find knowledge from large amounts of data. The methodical development of such methods will facilitate the campaign of human contribution for efficient and effective data examination.

8.8.10. Disseminated mining of data and real-time mining of data stream mining

Modern data mining techniques, developed to work at the centralized location, don't work well in several disseminated computing environments existent today. Developments in disseminated data mining techniques are expected. Furthermore, numerous applications comprising stream data need vibrant models of data mining to be made in real-time (Wong, 1999). Extra research is required in this particular direction.

8.8.11. Information security and protection of privacy in data mining

The richness of confidential or personal information accessible in the electronic forms joined with progressively powerful tools of data mining, poses the risk to data security. Increasing curiosity in the mining of data for counterterrorism also enhances the concern (Cooper & Collman, 2005)

REFERENCES

1. Aggarwal, P., & Chaturvedi, M. M. (2013). Application of data mining techniques for information security in a cloud: a survey. *International Journal of Computer Applications*, *80*(13), 11-17.
2. Allam, O., Gray, A., Bailey, H., & Morrey, D. (2006). Primary care oncology: addressing the challenges. *Informatics in primary care*, *14*(3), 5-10.
3. Bachlechner, D., Thalmann, S., & Maier, R. (2014). Security and compliance challenges in complex IT outsourcing arrangements: A multi-stakeholder perspective. *Computers & Security*, *40*, 38-59.
4. Bailey, H., & Urquhart, C. (2003). Information needs in a clinical trials unit. *Health Informatics Journal*, *9*(2), 111-126.
5. Beer, D. (2007). Thoughtful territories: imagining the thinking power of things and spaces. *City*, *11*(2), 229-238.
6. Bernal, J. L., Cummins, S., & Gasparrini, A. (2017). Interrupted time series regression for the evaluation of public health interventions: a tutorial. *International journal of epidemiology*, *46*(1), 348-355.
7. Bhaskaran, K., Gasparrini, A., Hajat, S., Smeeth, L., & Armstrong, B. (2013). Time series regression studies in environmental epidemiology. *International journal of epidemiology*, *42*(4), 1187-1195.
8. Bignami, F. (2007). European Versus American Liberty: A Comparative Privacy Analysis of Antiterrorism Data Mining. *BCL Rev.*, *48*, 609.
9. Borgavakar, S. P., & Shrivastava, A. (2017). Evaluating student's performance using k-means clustering. *Int. J. Eng. Res. Technol*, *6*(05), 114-116.
10. Borriello, G. (2000). The challenges to invisible computing. *Computer*, *33*(11), 123-125.
11. Borriello, G. (2008). Invisible computing: automatically using the many bits of data we create. *Philosophical Transactions of the Royal Society A: Mathematical, Physical and Engineering Sciences*, *366*(1881), 3669-3683.
12. Bryce, J., & Klang, M. (2009). Young people, disclosure of personal information and online privacy: Control, choice and consequences. *Information security technical report*, *14*(3), 160-166.
13. Cannataro, M., Congiusta, A., Pugliese, A., Talia, D., & Trunfio,

P. (2004). Distributed data mining on grids: services, tools, and applications. *IEEE Transactions on Systems, Man, and Cybernetics, Part B (Cybernetics)*, *34*(6), 2451-2465.

14. Chen, M., Ebert, D., Hagen, H., Laramee, R. S., Van Liere, R., Ma, K. L., ... & Silver, D. (2008). Data, information, and knowledge in visualization. *IEEE computer graphics and applications*, *29*(1), 12-19.
15. Chen, W., Huang, Z., Wu, F., Zhu, M., Guan, H., & Maciejewski, R. (2017). VAUD: A visual analysis approach for exploring spatio-temporal urban data. *IEEE transactions on visualization and computer graphics*, *24*(9), 2636-2648.
16. Combs, G. M. (2003). The duality of race and gender for managerial African American women: Implications of informal social networks on career advancement. *Human Resource Development Review*, *2*(4), 385-405.
17. Cooper, T., & Collman, J. (2005). Managing information security and privacy in healthcare data mining. *Medical Informatics*, 2(2), 95-137.
18. Cuéllar, M. F. (2002). The tenuous relationship between the fight against money laundering and the disruption of criminal finance. *J. Crim. L. & Criminology*, *93*, 311.
19. D'Oca, S., & Hong, T. (2015). Occupancy schedules learning process through a data mining framework. *Energy and Buildings*, *88*, 395-408.
20. Da Costa, A. F., & Manzato, M. G. (2016). Exploiting multimodal interactions in recommender systems with ensemble algorithms. *Information Systems*, *56*, 120-132.
21. Damle, C., & Yalcin, A. (2007). Flood prediction using time series data mining. *Journal of Hydrology*, *333*(2-4), 305-316.
22. Dietrich, G., Krebs, J., Fette, G., Ertl, M., Kaspar, M., Störk, S., & Puppe, F. (2018). Ad hoc information extraction for clinical data warehouses. *Methods of information in medicine*, *57*(1), 22-29.
23. Dimitoglou, G., Mendiboure, C., Reardon, K., & Sanchez-Duarte, L. (1998). Whole Sun Catalog: Design and Implementation. In *Three-Dimensional Structure of Solar Active Regions*, 3(1), 155, 297.
24. Dixit, K., & Pandya, B. (2014). An overview of Multiplicative data perturbation for privacy preserving Data mining. *International Journal for research in applied science and engineering technology (I JRAS ET)*, *2*, 90-96.

25. Eggert, K. (2011). Foreclosing on the Federal Power Grab: Dodd-Frank, Preemption, and the State Role in Mortgage Servicing Regulation. *Chap. L. Rev.*, *15*, 171.
26. Esling, P., & Agon, C. (2012). Time-series data mining. *ACM Computing Surveys (CSUR)*, *45*(1), 1-34.
27. Faloutsos, C., Ranganathan, M., & Manolopoulos, Y. (1994). Fast subsequence matching in time-series databases. *ACM Sigmod Record*, *23*(2), 419-429.
28. Fernández, A., Peralta, D., Benítez, J. M., & Herrera, F. (2014). E-learning and educational data mining in cloud computing: an overview. *International Journal of Learning Technology*, *9*(1), 25-52.
29. Gane, N., Venn, C., & Hand, M. (2007). Ubiquitous surveillance: interview with Katherine Hayles. *Theory, Culture & Society*, *24*(7-8), 349-358.
30. Ganguly, A. R., Kodra, E. A., Agrawal, A., Banerjee, A., Boriah, S., Chatterjee, S., ... & Wuebbles, D. (2014). Toward enhanced understanding and projections of climate extremes using physics-guided data mining techniques. *Nonlinear Processes in Geophysics*, *21*(4), 777-795.
31. Gholamian, M., Fathian, M., Julashokri, M., & Mehrbod, A. (2011). Improving electronic customers' profile in recommender systems using data mining techniques. *Management Science Letters*, *1*(4), 449-456.
32. Goebel, M., & Gruenwald, L. (1999). A survey of data mining and knowledge discovery software tools. *ACM SIGKDD explorations newsletter*, *1*(1), 20-33.
33. Golfarelli, M., Maio, D., & Rizzi, S. (1998). The dimensional fact model: A conceptual model for data warehouses. *International Journal of Cooperative Information Systems*, *7*(02n03), 215-247.
34. Gonzalez, H. B. (1996). New and Continuing Challenges in the Fight Against Money Laundering. *Fordham Int'l LJ*, *20*, 1543.
35. Gunay, B., & Shen, W. (2017). Connected and distributed sensing in buildings: Improving operation and maintenance. *IEEE Systems, Man, and Cybernetics Magazine*, *3*(4), 27-34.
36. Hormozi, A. M., & Giles, S. (2004). Data mining: A competitive weapon for banking and retail industries. *Information systems management*, *21*(2), 62-71.

37. Hossain, M. S., Muhammad, G., Abdul, W., Song, B., & Gupta, B. B. (2018). Cloud-assisted secure video transmission and sharing framework for smart cities. *Future Generation Computer Systems, 83*, 596-606.
38. Hossain, M. Z., Akhtar, M. N., Ahmad, R. B., & Rahman, M. (2019). A dynamic K-means clustering for data mining. *Indonesian Journal of Electrical engineering and computer science, 13*(2), 521-526.
39. Hurley, J., Morris, S., & Portelance, G. (2019). Examining the debt implications of the Belt and Road Initiative from a policy perspective. *Journal of Infrastructure, Policy and Development*, *3*(1), 139-175.
40. Indra, E. (2017). Aplikasi Pendataan Lokasi Bengkel Resmi Sepeda Motor Di Kota Medan Berbasis Android Menggunakan Algoritma Floyd Warshall. *Jurnal Keperawatan Priority*, *1*(1), 4-19.
41. Irfan, R., King, C. K., Grages, D., Ewen, S., Khan, S. U., Madani, S. A., ... & Li, H. (2015). A survey on text mining in social networks. *The Knowledge Engineering Review*, *30*(2), 157-170.
42. Jeba, N., & Rathi, S. (2021). Effective data management and real-time analytics in internet of things. *International Journal of Cloud Computing*, *10*(1-2), 112-128.
43. Kallio, A., Vuokko, N., Ojala, M., Haiminen, N., & Mannila, H. (2011). Randomization techniques for assessing the significance of gene periodicity results. *BMC bioinformatics*, *12*(1), 1-14.
44. Kim, K. J., Ahn, H., & Jeong, S. (2010). Context-aware recommender systems using data mining techniques. *International Journal of Industrial and Manufacturing Engineering*, *4*(4), 381-386.
45. Kirkos, E., Spathis, C., & Manolopoulos, Y. (2007). Data mining techniques for the detection of fraudulent financial statements. *Expert systems with applications*, *32*(4), 995-1003.
46. Koren, Y. (2010). Factor in the neighbors: Scalable and accurate collaborative filtering. *ACM Transactions on Knowledge Discovery from Data (TKDD)*, *4*(1), 1-24.
47. KRaja, D. R. An Approach to Improve Cloud Data Privacy by Preventing from Data Mining Based Attacks. *International Journal of Scientific and Research Publications*, 558, 5-10.

48. Kriegel, H. P., Borgwardt, K. M., Kröger, P., Pryakhin, A., Schubert, M., & Zimek, A. (2007). Future trends in data mining. *Data Mining and Knowledge Discovery*, *15*(1), 87-97.
49. Lakshmi, D. B., & Arundathi, S. (2014). Providing Privacy and Security for Cloud Data Using Data Mining. *International Journal of Innovation and Scientific Research, ISSN*, 3, 2351-8014.
50. Lin, J., Keogh, E., Wei, L., & Lonardi, S. (2007). Experiencing SAX: a novel symbolic representation of time series. *Data Mining and knowledge discovery*, *15*(2), 107-144.
51. Linden, G., Smith, B., & York, J. (2003). Amazon. com recommendations: Item-to-item collaborative filtering. *IEEE Internet computing*, *7*(1), 76-80.
52. Liu, L., Wang, S., Peng, Y., Huang, Z., Liu, M., & Hu, B. (2016). Mining intricate temporal rules for recognizing complex activities of daily living under uncertainty. *Pattern Recognition*, *60*, 1015-1028.
53. Mandala, I. G. N. N., Nawangpalupi, C. B., & Praktikto, F. R. (2012). Assessing credit risk: An application of data mining in a rural bank. *Procedia Economics and Finance*, *4*, 406-412.
54. Manikandan, V., Porkodi, V., Mohammed, A. S., & Sivaram, M. (2018). Privacy preserving data mining using threshold based fuzzy cmeans clustering. *ICTACT Journal on Soft Computing*, *9*(1), 3-10.
55. Marbán, O., Segovia, J., Menasalvas, E., & Fernández-Baizán, C. (2009). Toward data mining engineering: A software engineering approach. *Information systems*, *34*(1), 87-107.
56. Mattern, F., Ortega Cantero, M., & Lorés Vidal, J. (2001). Ubiquitous computing. *Internet@ Future, Jahrbuch Telekommunikation und Gesellschaft*, 1(2), 52-61.
57. Moreno, M. N., Segrera, S., López, V. F., Muñoz, M. D., & Sánchez, Á. L. (2016). Web mining based framework for solving usual problems in recommender systems. A case study for movies′ recommendation. *Neurocomputing*, *176*, 72-80.
58. Najafabadi, M. K., Mohamed, A. H., & Mahrin, M. N. R. (2019). A survey on data mining techniques in recommender systems. *Soft Computing*, *23*(2), 627-654.
59. Nakatsuji, M., Toda, H., Sawada, H., Zheng, J. G., & Hendler, J. A. (2016). Semantic sensitive tensor factorization. *Artificial Intelligence*, *230*, 224-245.

60. Ngai, E. W., Hu, Y., Wong, Y. H., Chen, Y., & Sun, X. (2011). The application of data mining techniques in financial fraud detection: A classification framework and an academic review of literature. *Decision support systems*, *50*(3), 559-569.
61. Nilashi, M. (2016). An overview of data mining techniques in recommender systems. *Journal of Soft Computing and Decision Support Systems*, *3*(6), 16-44.
62. Orlando, S., Orsini, R., Raffaetà, A., Roncato, A., & Silvestri, C. (2007). Trajectory data warehouses: Design and implementation issues. *Journal of computing science and engineering*, *1*(2), 211-232.
63. Pappalardo, L., Vanhoof, M., Gabrielli, L., Smoreda, Z., Pedreschi, D., & Giannotti, F. (2016). An analytical framework to nowcast well-being using mobile phone data. *International Journal of Data Science and Analytics*, *2*(1), 75-92.
64. Park, D. H., Kim, H. K., Choi, I. Y., & Kim, J. K. (2012). A literature review and classification of recommender systems research. *Expert systems with applications*, *39*(11), 10059-10072.
65. Park, D. H., Kim, H. K., Kim, J. K., Choi, I. Y., & Kim, J. K. (2011). A review and classification of recommender systems research. *International Proceedings of Economics Development & Research*, *5*(1), 290.
66. Patel, T., & Patel, V. (2020). Data privacy in construction industry by privacy-preserving data mining (PPDM) approach. *Asian Journal of Civil Engineering*, *21*(3), 505-515.
67. Pike, R., & Cheng, N. S. (2001). Credit management: an examination of policy choices, practices and late payment in UK companies. *Journal of Business Finance & Accounting*, *28*(7-8), 1013-1042.
68. Rajendran, K., Jayabalan, M., & Rana, M. E. (2017). A study on k-anonymity, l-diversity, and t-closeness techniques. *IJCSNS*, *17*(12), 172.
69. Ramageri, B. M., & Desai, B. L. (2013). Role of data mining in retail sector. *International Journal on Computer Science and Engineering*, *5*(1), 47.
70. Ravisankar, P., Ravi, V., Rao, G. R., & Bose, I. (2011). Detection of financial statement fraud and feature selection using data mining techniques. *Decision support systems*, *50*(2), 491-500.

71. Romero, C., & Ventura, S. (2013). Data mining in education. *Wiley Interdisciplinary Reviews: Data Mining and Knowledge Discovery*, *3*(1), 12-27.
72. Rosania, G. R., Crippen, G., Woolf, P., & Shedden, K. (2007). A cheminformatic toolkit for mining biomedical knowledge. *Pharmaceutical Research*, *24*(10), 1791-1802.
73. Sarasvathi, N., & Fong, J. Z. X. (2018). Study and Implementation of Internet of Things (IoT) Based Vehicle Safety Alert and Tracking System. *INTI Journal*, *1*(10), 1-11.
74. Sasireka, K., & Raja, K. (2014). An approach to improve cloud data privacy by preventing from data mining based attacks. *International Journal of Scientific and Research Publications*, *4*(2), 1-4.
75. Schermer, B. W. (2011). The limits of privacy in automated profiling and data mining. *Computer Law & Security Review*, *27*(1), 45-52.
76. Serra, J., & Arcos, J. L. (2014). An empirical evaluation of similarity measures for time series classification. *Knowledge-Based Systems*, *67*, 305-314.
77. Setoguchi, S., Schneeweiss, S., Brookhart, M. A., Glynn, R. J., & Cook, E. F. (2008). Evaluating uses of data mining techniques in propensity score estimation: a simulation study. *Pharmacoepidemiology and drug safety*, *17*(6), 546-555.
78. Sharma, S., Chugh, A., & Kumar, A. (2013). Enhancing data security in cloud storage. *International Journal of Advanced Research in Computer and Communication Engineering*, *2*(5), 2132-2134.
79. Shaw, M. J., Subramaniam, C., Tan, G. W., & Welge, M. E. (2001). Knowledge management and data mining for marketing. *Decision support systems*, *31*(1), 127-137.
80. Shneiderman, B. (2002). Inventing discovery tools: combining information visualization with data mining. *Information visualization*, *1*(1), 5-12.
81. Singh, S., & Sapra, R. (2014). Secure replication management in cloud storage. *International Journal of Emerging Trends and Technology in Computer Science*, *3*(2), 251-254.
82. Steinerberger, S. (2015). On the number of positions in chess without promotion. *International Journal of Game Theory*, *44*(3), 761-767.
83. Thies, F., Huber, A., Bock, C., Benlian, A., & Kraus, S. (2019). Following the crowd—does crowdfunding affect venture capitalists'

selection of entrepreneurial ventures?. *Journal of Small Business Management*, *57*(4), 1378-1398.

84. Twedt, D. W. (1952). A multiple factor analysis of advertising readership. *Journal of Applied Psychology*, *36*(3), 207.
85. Vaghashia, H., & Ganatra, A. (2015). A survey: privacy preservation techniques in data mining. *International Journal of Computer Applications*, *119*(4), 4-19.
86. Verykios, V. S., Bertino, E., Fovino, I. N., Provenza, L. P., Saygin, Y., & Theodoridis, Y. (2004). State-of-the-art in privacy preserving data mining. *ACM Sigmod Record*, *33*(1), 50-57.
87. Visser, H., & Molenaar, J. (1995). Trend estimation and regression analysis in climatological time series: an application of structural time series models and the Kalman filter. *Journal of Climate*, *8*(5), 969-979.
88. Wong, P. C. (1999). Visual data mining. *IEEE Computer Graphics and Applications*, *19*(5), 20-21.
89. Wu, X., Zhu, X., Wu, G. Q., & Ding, W. (2013). Data mining with big data. *IEEE transactions on knowledge and data engineering*, *26*(1), 97-107.
90. Xie, T., Thummalapenta, S., Lo, D., & Liu, C. (2009). Data mining for software engineering. *Computer*, *42*(8), 55-62.
91. Xu, L. D., & Duan, L. (2019). Big data for cyber physical systems in industry 4.0: a survey. *Enterprise Information Systems*, *13*(2), 148-169.
92. Xu, L., Jiang, C., Wang, J., Yuan, J., & Ren, Y. (2014). Information security in big data: privacy and data mining. *Ieee Access*, *2*, 1149-1176.
93. Yassine, A., Singh, S., Hossain, M. S., & Muhammad, G. (2019). IoT big data analytics for smart homes with fog and cloud computing. *Future Generation Computer Systems*, *91*, 563-573.
94. Zaiane, O. R., Han, J., Li, Z. N., & Hou, J. (1998). Mining multimedia data, 98, 83-96.
95. Zhang, Y., Qiu, M., Tsai, C. W., Hassan, M. M., & Alamri, A. (2015). Health-CPS: Healthcare cyber-physical system assisted by cloud and big data. *IEEE Systems Journal*, *11*(1), 88-95.

INDEX

Printed in the United States
by Baker & Taylor Publisher Services